G. Henrici-Olivé   S. Olivé

The Chemistry of the Catalyzed Hydrogenation of Carbon Monoxide

With 48 Figures

Springer-Verlag
Berlin Heidelberg NewYork Tokyo 1984

Dr. Gisela Henrici-Olivé
Prof. Dr. Salvador Olivé

University of California, San Diego, Department of Chemistry
La Jolla, CA 92037/USA

ISBN 3-540-13292-9 Springer-Verlag Berlin Heidelberg New York Tokyo
ISBN 0-387-13292-9 Springer-Verlag New York Heidelberg Berlin Tokyo

© Springer-Verlag Berlin Heidelberg 1984
Printed in Germany

Satz, Druck und Einband: Graphischer Betrieb, Konrad Triltsch, Würzburg
2154/3140-543210

Preface

During the oil embargo, in the winter 1973/74, parts of Western Europe present-ed an almost war-like aspect on Saturdays and Sundays: no traffic on the high-ways, no crowds at ski resorts and other weekend entertainment places, no gaso-line at the pumps.

Living and teaching then in that part of the world, and discussing the situa-tion with our students, we came to the conclusion that it would be timely to col-lect the fine chemistry already known at the time in the field of conversion of coal to gasoline and other chemicals, and by this way help to draw the attention to this important alternative to crude oil. The idea of this book was born.

The energy shock of the early seventies has been healthy and of great conse-quences in chemistry. Large amounts of research money have been put to work since, and our knowledge of the possibilities and limitations of coal-based chemistry has increased enormously. During several years it appeared inap-propriate to write a monograph about a topic which was in the midst of such an impetuous development. Nevertheless, we collected, and critically selected, the upcoming work as it appeared in the literature, and also tried to provide some modest input ourselves.

Now, ten years later, the situation seems to be settled to a certain degree. Partly, this is due to the fact that oil euphoria gains ground once more: oil prices are going down, and Governments, Big Business and other suppliers of research money tend to forget difficult situations rapidly. This has led to a reduced output of research papers in the field during the past one or two years. It should, however, be noticed that several individuals and institutions raise "warning flags". A number of estimates indicate that synfuel production will be substantial in the relatively near future, for despite the present glut of natural gas and oil, the underlying oil and gas situation that permitted the oil embargo is unchanged. A recent study of the American Institute of Chemical Engineers predicts the need for synthetic fuel in USA by about the year 2000, with genuine possibilities for its abrupt need earlier. Increasing awareness of these prospects may soon lead to a new boom in syngas research.

While synfuel technology still may have a far way to go to bring production costs to a competitive level, the chemistry of the use of syngas from coal ($CO + H_2$) for catalytic organic syntheses has reached a certain maturity, at least to a degree where "stock taking" may be a help for assessing the present state of the art and defining future research goals.

With this general background in mind we now venture to introduce our monograph "The Chemistry of the Catalyzed Hydrogenation of Carbon Monoxide".

La Jolla, 1984   G. Henrici-Olivé   S. Olivé

Contents

1 Introduction

The catalyzed hydrogenation and condensation of carbon monoxide to hydro-carbons and/or alcohols and other oxygenates have attracted much attention because of the need to develop alternatives to petroleum feedstocks. The origin of the development goes back to the pre-World War II Germany. Possessing large coal stocks, but completely inadequate oil resources, Germany was forced to undertake the transformation of coal into liquid hydrocarbons. As early as 1924, Franz Fischer published a book on this subject. In the following two decades, two processes were simultaneously developed to great industrial importance: the catalytic direct hydrogenation of coal at high pressure (*Bergius process*), and the catalytic normal to medium pressure synthesis of hydro-carbons from carbon monoxide and hydrogen (*Fischer-Tropsch process*). During World War II this development reached its zenith.

After 1945, the situation changed. The oil supply became plentiful, oil prices began to fall, and Industry was ready to forget good old coal. Until the Arab oil embargo in the early seventies, with escalating oil prices in its after-math, taught the Western World that cheap oil cannot be taken for granted. Apart from the search for other sources of energy, such as nuclear, solar, geothermal, water and wind power, scientific and industrial research reverted to coal, and a decade of intensive and fruitful work in the field of CO hydrogenation started.

At present, oil prices appear to have stabilized. However, apart from price considerations and the fear of future politically motivated oil embargos, there are good reasons for a consistent, continuous research in the field of coal based chemistry. The rising curve of demand will approach, and eventually cross, the falling curve of world wide crude oil supply. When that era arrives, Industry should be prepared to move into it with proved technology and in-place facilities. A number of individuals and institutions do not rule out a change in the energy situation in the near future.

Wender (1983), in a Plenary Lecture to the International Conference on Coal Science, cited a study of the American Institute of Chemical Engineers predicting the need for synthetic fuels in USA by about the year 2000, "with genuine possibilities for its abrupt need earlier". Another study, by Cambridge Energy Associates (1983), predicts that even a modest world recovery could bring about an explosion in demand for OPEC oil, and the next oil crisis as early as 1986. The International Energy Agency (1982), an organization of 21 nations formed to monitor energy trends and allocate oil in the event of supply disruptions, took a similar view. Also Greene (1983), Stanford Research Institute in Menlo Park, California, warned that by the end of the decade "a significant liquid fuel crunch may come back and hit us very hard".

The exploration of new methods that promise to open up more coal reserves are well underway. Underground coal gasification might become an alternative to classical coal mining with its declining appeal, its inherent danger, the escalating price of extraction, and the somewhat limited reserves available to the existing methods (Haggin, 1983). Moreover, coal mined according to classical procedures needs to be gasified to syngas:

$$C_{(coal)} + H_2O_{(1)} \rightarrow CO_{(g)} + H_{2(g)} \qquad - 176 \text{ kJ/mol}.$$

The reaction, as indicated, is *endothermic* and it proceeds with interesting rates at $T > 800\,°C$. Underground coal gasification leads directly to carbon monoxide according to

$$C_{(coal)} + 1/2\ O_{2(g)} \rightarrow CO_{(g)} \qquad + 109 \text{ kJ/mol}.$$

This reaction is *exothermic* and, therefore, can proceed at lower temperature. Of course, there are many other reactions taking place, and feed gas from an underground coal gasification burn needs removal of byproducts. Obviously, the reaction has to be carefully controlled to prevent excessive losses, e.g. as CO_2. After adjustment of the H_2/CO ratio, the syngas could, in principle, be fed directly to a Fischer-Tropsch or Mobil gasoline plant. Another approach to improved coal conversion is the application of supercritical fluid techniques (Worthy, 1983).

Production costs of synfuel from CO and H_2 can be expected to fall as technology improves. As an example, the Sasol-3 synthetic fuel plant (with Sasol-2 and Sasol-1 in Sasolburg, South Africa, the only large scale Fischer-Tropsch synthesis plants presently in operation) costs $ 1 billion less than did Sasol-2 (Wender, 1983).

Chemical research needs to keep up with the developments. It has already gone a long way since the early days of the Fischer-Tropsch synthesis, when heterogeneous catalysis was seen from the point of view of "active surfaces", with "physisorption" and "chemisorption" and "desorption" of reactants and products as the dominating conceptions. It was the advent of modern coordination chemistry, and the increasing understanding of *homogeneous* catalytic processes on transition metal compounds, which led to a new view of *heterogeneous* catalysis, based on individual metal centers and their coordination chemistry.

Similar development processes have taken place in other branches of chemistry. Thus, Colloid Chemistry has been plagued for almost a century by inadequate experiments and theories, until Matijević started to see the phenomena on a molecular base, and introduced the concept of complex formation for the understanding of generation and stabilization of colloidal particles (Matijević, 1973). Ziegler polymerization of ethylene with transition metal compound/aluminum alkyl catalysts is another example (Ziegler et al., 1955). Ziegler thought of the aluminum alkyl as the catalyst, and attributed a somewhat undefined role as "cocatalyst" to the transition metal component. For many years, several schools of thought maintained the firm belief that chain growth takes place at a base metal-carbon bond. However, as the experiences of inorganic coordination chemistry, as well as quantum mechanical considerations, found their way into the thinking of polymer chemists, it became obvious that the transition metal with its d orbitals available for σ and π bonding is the true catalytic center, whereas the aluminum alkyl fulfils a number of cocatalyst tasks. (For a review see Henrici-Olivé and Olivé, 1981). Fuel cells have been the sacred province for electrochemists. A fuel cell is efficient because it can operate at isothermal conditions; catalyst chemists are the ones who can make that happen (Haensel, 1978).

The past decade has seen a forceful research impetus in many laboratories, in different countries. Scientists of many disciplines (organic, inorganic, coordination chemists, spectroscopists, crystallographers, surface scientists, physicists) have tried to contribute to the elucidation of the reaction mechanisms involved in the catalytic CO hydrogenation, as well as in the development of new syntheses (e.g. methanol, glycol and higher polyalcohols from $CO + H_2$; acetic acid from methanol + CO; hydrocarbons from methanol). Not always has this multilateral approach been easy, or even fruitful. Thus, much work has been invested in the *ultraviolet photoelectron spectroscopy* of carbon monoxide on various metal surfaces, before it became clear that no meaningful information could be obtained. For instance, the spectra of CO on Ru, Fe, Ni, Co, Rh, Pd, Ir are all the like (e.g. Bonzel and Fischer, 1975; Conrad et al., 1976), whereas the catalytic activity of these metals, under comparable conditions, varies by a factor of 100 (Vannice, 1975). Apparently, this form of spectroscopy is insensitive to the details affecting the chemistry most profoundly (Moskovits, 1979).

Another example of misunderstanding between different groups of workers refers to the elucidation of the promoter effect of *potassium* in the Fischer-Tropsch synthesis with iron catalysts. Surface scientists and physicists have studied the changes of adsorption energy and IR spectra of CO on iron in the absence and presence of metallic potassium, and have drawn conclusions as to changes in CO orbital energies and bonding modes and their effects on catalytic activity (e.g. Brodén et al. 1979; Garfunkel et al., 1982). These studies have not taken into account that the conditions of the Fischer-Tropsch synthesis (presence of H_2O, CO_2) are *incompatible with the presence of metallic potassium;* the promoter is certainly present as K^+, even if it would have been transformed to metallic potassium during the reductive activation of the catalyst (for a review see Henrici-Olivé and Olivé, 1982).

Another dispute among the various factions of the interested scientific community refers to the way of *C–C bond formation* in the Fischer-Tropsch synthesis. Some advocate the "carbide theory" put forward by Fischer and Tropsch themselves in the twenties, involving the formation of CH_x entities from metal carbide on the catalytic surface, and their growing together to large hydrocarbon molecules; others prefer the CO insertion mechanism, in line with coordination chemistry and homgeneous catalysis. (For a discussion of this question, at the very heart of the mechanistic aspects of the Fischer-Tropsch synthesis, see Chap. 9).

We have tried to put the different opinions into perspective, and provide an overview of the present state of catalytic CO hydrogenation, as far as its chemistry is concerned. We have deliberately excluded the technological and economical aspects, since excellent books are available covering these fields, to which the reader is referred (Storch et al., 1951, for coverage of the early work; Falbe, 1977, 1980).

After these introductory remarks it will be evident that our treatment of the chemistry of the catalyzed CO hydrogenation will emphasize processes taking place in the coordination sphere of the catalytic metal centers, either in *metal surfaces* (heterogeneous catalysis) or in *molecular complexes* (homogeneous catalysis). In this sense we considered it useful to initiate the book with a basic treatment of the interactions of hydrogen and carbon monoxide with such metal centers (Chap. 2–4). Chapter 5, treating in a very general way with the key reactions making up a catalytic cycle, is considered of particular significance.

It should be noted in this context that one important role of organometallic and coordination chemistry is that of providing stable models for species postulated as intermediates, thus permitting an investigation of their chemistry when bonded to a transition metal center. It may also be noted that the reactivity of isostructural transition metal complexes frequently follows the order 1st row > 2nd row > 3rd row. This has the important consequence that 3rd row complexes often react too slowly for being effective as catalysts; however, they are sometimes extremely useful for modelling intermediates in catalytic cycles.

In these first Chapters, as well as throughout the book, it is our explicit intention to emphasize the analogies between homogeneous and heterogeneous reaction patterns, rather than to strictly treat them separately. Of course, certain inherent differences must be taken into account.

After some considerations about catalysts, supports, and their interactions in Chap. 6, the treatment of the catalytic CO hydrogenation proper starts with Chap. 7. *Methanation*, as the most simple of the hydrogenation reactions, is particularly suited for mechanistic studies (Chap. 7). The same is true, to a certain degree, with the *methanol synthesis* (Chap. 8). *Fischer-Tropsch synthesis* (Chap. 9), is probably the most intensively investigated of the CO hydrogenation reactions up to-day. As a "reductive polymerization" of CO, it follows polymerization statistics, and leads to a wide variety of molecular weights; moreover the reaction products (hydrocarbons and/or oxygenates) depend on the reaction conditions. Therefore, *selectivity* (or the lack thereof) is a major problem. Because of the high selectivity often associated with homogeneous catalytic reactions, they may appear to be more appropriate for the selective production of chemicals (Chap. 10). To a certain degree this is true for the hydroformylation of olefins and formaldehyde, if the principles of homogeneous catalysis (in particular ligand influences) are properly taken care of. The production of polyalcohols, however, is also hampered by a lack of selectivity.

As Herman Bloch has once remarked: everything that can happen in a catalytic reaction will happen; selectivity is just a matter of relative rates (cited by Haensel, 1978).

Chapter 11 is concerned with *methanol* as a raw material. Methanol from $CO + H_2$ is by now an amply available commodity product, and its use in catalysis enjoys ever increasing importance (e.g. acetic acid synthesis, Mobil process for synfuel production). In the last Chapter, we try to give a unified overview of the various *CO hydrogenation mechanisms*.

The literature up to mid 1983 has been considered. It should, however, be noted, that an exhaustive review of all work done in this vast field is plainly impossible, and has not been intended. Inevitably, the selection has to be somewhat subjective. Where possible more recent papers are cited, which may guide the reader to earlier work of the same or other authors. Although literature reports have amply been consulted, and are duly referenced, the mechanistic implications given in this book represent, in many instances, the authors' present opinion.

One final remark may be appropriate, concerning the *use of SI units*. As pointed out by Cotton (1980), there are some objections to be made to many aspects of the SI units, and the adoption of SI "in toto" and uncritically has not been shown to be either necessary or desirable. In this book we use some (e.g. Joules instead of calories, pm instead of Å), but not others (e.g. we retain atmospheres and Torr, as well as degree Celsius).

1.1 References

Bonzel HP, Fischer TE (1975) Surface Sci 51:213
Brodén G, Gafner G, Bonzel HP (1979) Surface Sci 84:295
Cambridge, Energy Associates (1983) cited in Chemical Week, Sept. 7, p 37
Conrad H, Ertl G, Knözinger H, Küppers J, Latta EE (1976) Chem Phys Lett 42:115
Cotton FA, Wilkinson G (1980) Advanced Inorganic Chemistry. Interscience, 4th ed
Falbe J (ed) (1977) Chemierohstoffe aus Kohle. Georg Thieme, Stuttgart
Falbe J (ed) (1980) New Syntheses with Carbon Monoxide. Springer Heidelberg New York
Fischer F (1924) Herstellung flüssiger Kraftstoffe aus Kohle. Bornträger, Berlin
Garfunkel EL, Crowell JE, Somorjai GA (1982) J Phys Chem 86:310
Greene CR (1983) SRI International, Menlo Park, Calif, cited in Chemical Week, Sept. 7, 1983
Haensel V (1978) CHEMTECH, 414
Haggin J (1983) C & EN, July 18, p 15
Henrici-Olivé G, Olivé S (1981) CHEMTECH, 746
Henrici-Olivé G, Olivé S (1982) J Mol Catal 16:187
International Energy Agency (1982) cited in Chemical Week, Sept 7, p 37
Matijević E (1973) J Coll Interface Sci 43:217, and earlier references therein
Moskovits M (1979) Acc Chem Res 12:229
Storch HH, Golumbic N, Anderson RB (1951) The Fischer-Tropsch and Related Syntheses. Wiley, New York
Vannice MA (1975) J Catal 37:449, 462
Wender I (1983) cited in C & EN, August 22, p 7
Worthy W (1983) C & EN, August 29, p 18
Ziegler K, Holzkamp E, Breil H, Martin H (1955) Angew Chem 67:541

2 Transition Metal-Hydrogen Interactions

2.1 Reaction of Hydrogen with Transition Metal Complexes and Surfaces

In the early sixties, Vaska and Deluzio discovered that a well defined, soluble square planar iridium (I) complex can reversibly take up one molecule of hydrogen:

$$\text{trans-}[IrCl(CO)((C_6H_5)_3P)_2] + H_2 \rightleftarrows [H_2IrCl(CO)((C_6H_5)_3P)_2]. \qquad (2.1)$$

The hydrogen in this and similar complexes is activated, and these systems act as catalysts for the hydrogenation of olefins and other related reactions. (For a review see Vaska and Werneke, 1971.) From NMR and IR spectra of the complexes in chloroform solution, it was concluded that the hydrogen atoms occupy cis positions in the hydrogenated complex, and that the CO and Cl ligands are pushed out of their positions in the square plane, into approximately octahedral positions, with a Cl−Ir−CO angle of ca. 90° (see Fig. 2.1). Since iridium is capable of forming square planar (4 ligands), as well as

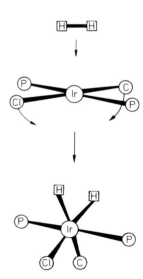

Fig. 2.1. Stereochemistry of hydrogen addition to square planar trans-[IrCl(CO)((C_6H_5)_3P)_2], according to Vaska and Werneke (1971)

octahedral (6 ligands) complexes, there are potentially free coordination sites available in the starting complex, which are then occupied by the two hydrogens in the resulting octahedral configuration.

From kinetic and thermodynamic studies, in polar and non-polar solvents, Schmidt et al. (1974) concluded that the addition of hydrogen to the Vaska complex and related compounds is a concerted process, with homolytic splitting of the H−H bond and simultaneous formation of the two metal−H bonds. The new bonds are essentially covalent σ bonds (see, however, Sect. 2.4 concerning a certain polarity of these bonds). The oxidation state of the metal center has changed in the course of the addition reaction from Ir(I) in the square planar, to Ir(III) in the octahedral complex. The term "oxidative addition" is used for this type of reaction (cf. Sect. 5.1).

The oxidative addition of hydrogen to a transition metal center has been observed since in many cases, and it plays a very important role in homogeneous catalysis (see e.g. Henrici-Olivé and Olivé, 1977). Not always do both hydrogen atoms of a H_2 molecule end up on the same metal center. In the reaction of $CO_2(CO)_8$ with hydrogen, for instance, two Co−H species are formed according to the overall equation:

$$Co_2(Co)_8 + H_2 \rightleftarrows 2\,HCo(CO)_4 \qquad\qquad (2.2)$$

although the detailed mechanism of this dissociative process is not known.

On the other hand, the interaction of molecular hydrogen with metal surfaces, as well as the catalytic activation of H_2 as a consequence of this interaction, are known since well over a century (Döbereiner, 1836), and many important industrial processes, such as for instance *hydrogenations*, the *Haber-Bosch ammonia synthesis*, and also the *Fischer-Tropsch synthesis* of hydrocarbons from CO and H_2, depend on it. Nevertheless, the nature of "adsorbed", or "chemisorbed" hydrogen appears to be still a subject of active research, as well as of different interpretation.

From a chemist's point of view, surface atoms on a metal catalyst fulfil the essential condition for an oxidative addition of hydrogen molecules. In contrast to the mass of the metal, where each metal atom is surrounded in all directions by other metal atoms, surface atoms offer free coordination sites. Moreover, it has been clearly established by H_2/D_2 exchange experiments that the adsorption of hydrogen, for instance on platinum, nickel or iridium single crystal surfaces, is dissociative (see Christman et al., 1976, 1979; Ibbotson et al., 1980). Thus, adopting an atomistic view, we can, in first approximation, equalize hydrogen adsorption on a transition metal surface with an oxidative addition of H_2 to a transition metal complex. For the time being, it may remain open whether the hydrogen atoms are taken up by the same metal atom [M(0) → M(II)], or by neighboring metal atoms [2 M(0) → 2 M(I)].

As discussed in the next few Sections, metal−H bonds in transition metal complexes and on metal surfaces have quite similar energetics, bond lengths, etc., which fact appears to corroborate this view.

2.2 Metal-Hydrogen Bond Energy

The driving force for the addition of hydrogen to a metal is related to the energy released during the formation of the two metal−hydrogen bonds. In first approximation, the bond energy, D_{M-H}, can be estimated from Eq. 2.3:

$$- \Delta H^0 = 2 D_{M-H} - D_{H-H} \tag{2.3}$$

where ΔH^0 is the enthalpy of the − exothermic − addition of hydrogen which can be experimentally determined; D_{H-H} is the dissociation energy of H_2, which is 431.4 kJ/mol. Equation 2.3 ignores the so-called reorganization energy, which represents the energy associated with the electronic and stereochemical rearrangement taking place during the addition; its magnitude is not exactly known in most cases, but it is generally assumed to be negligible, at least for a first approximation. In the particular case of the Vaska complex (Eq. 2.1),

$$\Delta H^0 = - 59 \text{ kJ/mol}.$$

From Eq. 2.3 follows then a bond energy

$$D_{Ir-H} = 245 \text{ kJ/mol}.$$

The bond energies for eleven related Ir−H complexes (Vaska and Werneke, 1971), as well as for $[CoH(CN)_5]^{3-}$ (De Vries, 1962) and $[CoH(CO)_4]$ (Alemdaroglu et al., 1976), have been found all to be in the range 240−290 kJ/mol.

Bond energies for atomic hydrogen on transition metal surfaces have been compiled by Martin (1978). These data have been obtained mainly from thermal desorption spectroscopy on single crystals. It turns out that − at least for the right half portions of the transition series − D_{M-H} is approximately constant at about 265 kJ/mol. Interestingly, the bond energy does not depend on the crystallographic plane; measurements on (100), (110) and (111) planes fall into the same range. Metal atoms located at edges on a surface could be expected to have different electronic properties and hence show different bond energies. but Christmann and Ertl (1976) showed that the difference of Pt−H bond energy on a flat Pt(111) surface, and on a stepped (997) surface does not exceed 2−3%. In accordance with this, also polycrystalline materials have M−H bond energies in the same range (Martin, 1978).

These findings suggest that a localized description of the valence electrons of the metal substrate and of the metal−H bond at a surface is appropriate. As in isolated transition metal complexes, *d*, *s*, and *p* orbitals of the metal may participate in the bonding ("spd hybrid orbitals"). In Ni−H bonds on metal surfaces, the 4*s* orbital appears to play the most important role, whereas for Pd−H and Pt−H, for instance, the 4*d* and 5*d* orbitals, respectively, are dominant. This has been made probable by theoretical studies (molecular orbital models for hydrogen adsorption, Fassaert et al., 1972; Fassaert and van der Avoird, 1976; Messmer et al., 1977), and experimentally shown by Demuth (1977), who studied photoemission spectra for hydrogen on the surfaces of polycrystalline Ni, Pd and Pt.

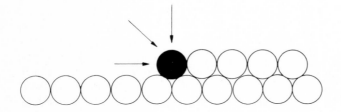

Fig. 2.2. Schematic model for metal atoms in the vicinity of a step in an otherwise flat surface

Evidently, the interaction of the surface metal atom with the underlying and surrounding metal atoms is not zero, and hence has to be considered. But qualitatively, this is comparable to the situation in transition metal hydride complexes, where the central metal atom is bonded to a number of other ligands. In the Vaska complex, for instance, these are one carbon monoxide, one chlorine and two phosphine molecules (cf. Fig. 2.1.). The other ligands can have a considerable influence on the reactivity and specificity of one particular M−H or M−C bond, through electronic as well as steric effects (see e.g. Henrici-Olivé and Olivé, 1977). Quantitatively, it may be a particular aspect of heterogeneous metal catalysts that most of the ligands of the metal center of a surface complex are other metal atoms, which is not generally the case in homogeneous transition metal catalysts (except, to a certain degree, for metal cluster compounds). The *d* electron energy level for a metal surface is in general higher than for molecular species (see e.g. Shustorovich et al., 1983). This, evidently, enhances orbital overlap between a relevant filled *d* orbital of a surface metal atom with the empty, antibonding σ^* orbital of a coordinated hydrogen molecule, facilitating electron flow from the metal into σ^*, thereby promoting the formation of the hydride surface complex.

The number of nearest neighbor metal atoms in a surface complex seems to play an important role in kinetic phenomena. Christmann and Ertl (1976) have found that a "step density" of 11% of the surface atoms, on an otherwise flat surface, increases the activity for H_2/D_2 exchange on such a surface by an order of magnitude. One evident difference between metal atoms located at a step and those within a flat surface is the availability of additional coordination sites on the former (see Fig. 2.2). Thus, although a M−H bond on a step does not appear to be much weaker than the others, a D_2 molecule (or any other suitable substrate molecule) can more easily become coordinatively bonded to such a step metal hydride, prior to the exchange (or any other) reaction.

2.3 Different Types of Metal-Hydrogen Bonding and the M−H Bond Length

The bond length of metal-hydrogen bonds in defined transition metal hydride complexes has been determined by X-ray single crystal structure analysis as well as by neutron diffraction. The X-ray diffraction is handicapped somewhat by the fact that the scattering of the X-rays is roughly proportional in amplitude to the number of electrons in the atom studied. Thus, the scattering of hydrogens tends to be swamped by that of the heavy metal atoms to which they are bonded. To improve accuracy, it was necessary to turn to neutron diffraction. Although the technique and theory of the two scattering methods are similar, the fundamental difference is that hydrogen atoms scatter neutrons almost as efficiently as most other elements. Inconvenients are the necessity of using large single crystals ($\simeq$ 10 mm^3), and the fact that a neutron source is not available on a routine basis.

Numerous transition metal hydride complexes have been investigated by neutron diffraction. The metal−H bond lengths given in Table 2.1 are taken

Table 2.1. M−H bond length determined by single crystal neutron diffraction (Bau et al., 1979).

Compound	M	M−H (pm)
	Terminal M−H Bonds	
$HMn(CO)_5$	Mn	160.1
$H_2Mo(C_5H_5)_2$	Mo	168.5
$H_3Ta(C_5H_5)_2$	Ta	177.4
$H_8Re_2(PEt_2Ph)_4$	Re	166.9
$H_4Os(PMe_2Ph)_3$	Os	165.9
	Bridging M−H−M Bonds	
$[HCr_2(CO)_{10}]^-[Et_4N]^+$	Cr	173.7
$HMo_2(C_5H_5)_2(CO)_4(PMe_2)$	Mo	185.1
$[HW_2(CO)_{10}]^-[Et_4N]^+$	W	171.8
$H_2Os_3(CO)_{10}$	Os	184.5
$H_8Re_2(PEt_2Ph)_4$	Re	187.8
$[H_3Ir_2(C_5Me_5)_2]^+BF_4^-$	Ir	175.0
$H_4Th_2(C_5Me_5)_4$	Th	229.0
	Triply Bridging M$_3$H Bonds	
$HFeCo_3(CO)_9(P(OMe)_3)_3$	Co	173.4
$H_3Ni_4(C_5H_5)_4$	Ni	169.1
	Interstitial Hydrides	
$[HCo_6(CO)_{15}]^-[(Ph_3P)_2N]^+$	Co	182.4
$[H_2Ni_{12}(CO)_{21}]^{2-}[(Ph_3P)_2N]_2^+$	Ni	184.0 − 200.0

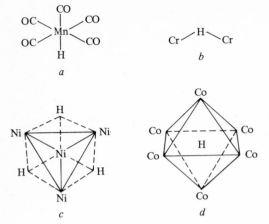

Fig. 2.3. Different types of metal-hydrogen bonding in transition metal complexes (according to Bau et al., 1979); a) terminal M−H bond in HMn(CO)$_5$; b) bridging M−H−M bond in the [HCr$_2$(CO)$_{10}$]$^-$ anion; c) triply bridging M$_3$H in the H$_3$Ni$_4$ core of H$_3$Ni$_4$(C$_5$H$_5$)$_4$; d) "interstitial" hydrogen: the HCo$_6$ core of the [HCo$_6$(CO)$_{15}$]$^-$ anion (all other ligands omitted for clarity)

from a compilation by Bau et al. (1979). Apart from the compounds with terminal M−H bonds (see Fig. 2.3a), complexes have been found where the hydrogen is simultaneously bonded to two metal centers, such as in [HCr$_2$(CO)$_{10}$]$^-$[Et$_4$N]$^+$. The metal−H−metal arrangement is in general somewhat bent (as indicated in Fig. 2.3b). The electron deficient three-center-two-electron bridge bonds are on average slightly longer than the terminal M−H bonds. In a few rare cases of cluster compounds, hydrogen has been found simultaneously bonded to three centers as, for instance, in H$_3$Ni$_4$(C$_5$H$_5$)$_4$, where the H$_3$Ni$_4$ core ressembles a cube with a missing corner (see Fig. 2.3c). Bau et al. (1979) suggested that this type of bonding might serve as a model for the bonding of hydrogen on metal surfaces, since the metal-metal distances associated with the H-capped faces are similar to those in the metal themselves.

In a couple of other clusters, a hydrogen was found right in the middle of an octahedral metal arrangement (Fig. 2.3d). The "interstitial" hydrogen has very nearly the same distance to each of the surrounding metal centers.

One final hydride structure appears noteworthy. Bau et al. (1981) investigated (by X-ray structural analysis) the complex FeH$_6$Mg$_4$Br$_{3.5}$Cl$_{0.5}$(THF)$_8$. The most remarkable feature of this compound is the [FeH$_6$]$^{4-}$ central core which shows an octahedral array of 6 H atoms with the Fe atom in the center (Fig. 2.4a). The average Fe−H bond length is 169 pm. Face capping Mg ions are disposed around this core in a tetrahedral fashion (Fig. 2.4b). One halide and the oxygen atoms of two THF molecules complete the coordination sphere around each Mg ion.

As to the metal-hydrogen bonds, it appears that most of the M−H bond lengths determined thus far fall into the relatively narrow range from 160 to 190 pm. This means that the hydrogen is situated generally close to what corresponds to a normal covalent bond.

If we turn now from the well defined molecular transition metal hydrido complexes to hydrogen adsorbed at metal surfaces, we have little material available for direct comparison. There is no way yet to measure M−H

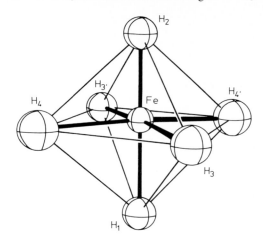

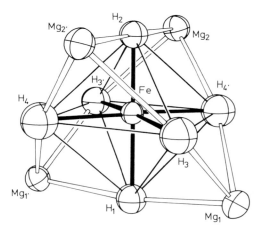

Fig. 2.4. Structure of the complex $FeH_6Mg_4Br_{3.5}Cl_{0.5}(THF)_8$; a) the $[FeH_6]^{4-}$ core; b) $[FeH_6]^{4-}$ surrounded by a tetrahedral array of magnesium ions, according to Bau et al., 1981 (THF ligands omitted for clarity). Reproduced by permission of the American Chemical Society

distances directly on a surface. However, sophisticated methods have been used to study the geometry of the hydrogen overlayer, as well as the surface coverage, on single crystals. In particular, low-energy-electron-diffraction (LEED), in combination with thermal desorption spectroscopy (TDS) and work function ($\Delta\varphi$) measurements have been applied to evaluate the geometry of metal−H bonding sites and to estimate the M−H bond length (Christmann et al., 1979). A trial-and-error procedure was necessary to fit experimental data and theory. In the case of the Ni(111) surface, Christmann et al. came to the conclusion that the most probable bonding mode is with hydrogen atoms occupying "threefold hollow adsorption sites", i.e. the hydrogens sit above holes in the cubic close-packed nickel surface at equal distance to three surface Ni atoms. With this supposition, a Ni−H bond length of 184 pm resulted from these studies. Evidently, this is the same type of bonding discussed above in context with the triple bridging M_3H bonds in well-defined complexes.

In an analysis of the metal orbitals associated with the (111) faces of cubic metals, Bond (1966) pointed out that there are no metal orbitals oriented normal to the surface. Instead, the orbitals are directed at points over the M_3 triangles. Apparently, these bits of information fit together and give a certain probability to the threefold bonding, at least for pure (111) surfaces. However, there is other theoretical work (Fassaert and van der Avoird, 1976) that, based on extended *Hückel molecular orbital calculations* predicts the on-top position as the most favorable one, followed by twofold bridges and threefold bridges (centered over threefold holes), and a decreasing stability of the metal−H bonding in the series (110) > (100) > (111), for nickel.

Summarizing this Section: *Thus far there is no discrepancy between the defined molecular complexes and metal surfaces with regard to the M−H bond length, but there seems to be a need for further work concerning the exact geometries of metal-hydrogen bonding on pertinent metal surfaces.*

2.4 Polarity of the M−H Bond

The observed M−H bond distances indicate that these bonds are normal covalent bonds. Nevertheless, they should be considered as somewhat polarized, and there is some confusion about the actual direction of the polarity (Henrici-Olivé and Olivé, 1976, and references therein). In general, it is assumed that the hydrogen has a partial negative charge, as borne out by the expressions hydride-, or hydrido-transition metal complex. The most frequently cited evidence for a negative charge on the hydrogen is the large high field shift in the nuclear magnetic resonance spectra of hydrogen directly bonded to a transition metal, in diamagnetic complexes (τ between 10 and 50 ppm). This shift is even used as a criterion for the presence of a hydrido-complex, particularly for species obtainable in solution only. Moreover, Chen et al. (1979) have concluded from X-ray photoelectron spectroscopic data of $HCo(CO)_4$, $HMn(CO)_5$ and $H_2Fe(CO)_4$ that the hydrogen atoms in these complexes are negatively charged, with approximate charges of − 0.75, − 0.8, and − 0.3 respectively. Concordantly, a molecular orbital calculation by Edgell and Gallup (1956) of the charge densities in $HCo(CO)_4$ indicates that 1.6 electrons are associated with the hydrogen atom, corresponding to a negative charge of 0.6. In aprotic solvents, the hydride ligand of some hydridocarbonyl complexes is sufficiently negative to react with Lewis acids such as BCl_3 or $AlCl_3$, with cleavage of the M−H bond. Examples are $HMn(CO)_5$, $HMn(CO)_4PPh_3$ and $HRe(CO)_5$.

On the other hand, hydrido-transition metal carbonyl complexes such as $HCo(CO)_4$ and related compounds, are slightly soluble in water, and such solutions are clearly acidic. The compound $HCo(CO)_4$ also forms adducts with nitrogen bases such as trialkylamines or pyridine. Calderazzo et al. (1981) deduced from X-ray single crystal analysis of the Et_3N adduct that the structure is best described as $Et_3NH^+...[Co(CO)_4]^-$, with the Co−H bond elongated from 0.156 nm in $HCo(CO)_4$ to 0.285 nm in the adduct. The interaction persists in hydrocarbon solution, and the protonic character of the nitrogen-bonded hydrogen is evidenced by a low field shift in ^{1}H-NMR.

Apparently, an acidic dissociation of a hydrido-transition metal complex is feasible only if the negative charge on the anionic part can be dissipated over several strongly electron accepting ligands, in particular CO. Moreover, the energy gain caused by the solvation of the H^+ ion in the case of the water solution, or by the formation of the protonated amine, will promote the reversal of the polarization of the M–H bond.

Even in solvents less polar than water, such as acetonitrile and methanol, certain hydridocarbonyl metal complexes are moderately acidic. Table 2.2 shows some examples. The acidity of the group 6B hydrides (Nrs. 1–3) in CH_3CN falls between that of CH_3SO_3H and CH_3COOH (in the same solvent), and decreases down the column of the Periodic Table. The influence of the polarity of the solvent is evident from Nrs. 4 and 12; the dissociation constant K_a is more than five orders of magnitude higher in methanol than in acetonitrile. $Ru_4(CO)_{13}H_2$ (two bridging carbonyls, Nr. 9) is somewhat more acidic than $Ru_4(CO)_{12}H_4$ (no bridging carbonyls, Nr. 10), indicating that bridging carbonyls are more electron withdrawing than terminal carbonyls. The pseudotetrahedral $Os_4(CO)_{12}H_4$, featuring bridging hydrides (Nr. 13), is considerably more acidic than the open chain trinuclear complex (Nr. 14) and the mononuclear species (Nr. 12), both having terminal hydride ligands only. This appears to indicate that for analogous complexes of the same element, bridging hydrides are more acidic (less basic) than the terminal ones (Walker et al., 1983). Though most of these complexes have to be considered as very weak acids (for comparison: the pK_s value for the autoionization of the solvent

Table 2.2. Acidity (pK_a values) of metal carbonyl hydrides in polar solvents; room temperature.

Nr.	Complex	Solvent	pK_a	Reference[a]
1	$(\eta^5\text{-}C_5H_5)Cr(CO)_3H$	CH_3CN	13.3	1
2	$(\eta^5\text{-}C_5H_5)Mo(CO)_3H$	CH_3CN	13.9	1
3	$(\eta^5\text{-}C_5H_5)W(CO)_3H$	CH_3CN	16.1	1
4	$Os(CO)_4H_2$	CH_3CN	20.8	1
5	$Os(CO)_4(CH_3)H$	CH_3CN	23.0	1
6	CH_3SO_3H	CH_3CN	10.0	1
7	CH_3COOH	CH_3CN	22.3	1
8	$Fe(CO)_4H_2$	CH_3OH	6.9	2
9	$Ru_4(CO)_{13}H_2$	CH_3OH	11.1	2
10	$Ru_4(CO)_{12}H_4$	CH_3OH	11.7	2
11	$Ru_4(CO)_{11}(P(OCH_3)_3)H_4$	CH_3OH	14.7	2
12	$Os(CO)_4H_2$	CH_3OH	15.2	2
13	$Os_4(CO)_{12}H_4$	CH_3OH	12.0	2
14	$Os_3(CO)_{12}H_2$	CH_3OH	14.4	2
15	CH_3COOH	CH_3OH	9.6	2

[a] 1. Jordan and Norton, 1982
2. Pearson and Ford, 1982

anhydrous methanol is 16.7), the acidity of the iron hydridocarbonyl complex (Nr. 8) is surprisingly high.

Typically, the rates of deprotonation (e.g. by methoxide ions) and of reprotonation (by methanol) are orders of magnitude slower for this type of complexes than the rates observed for similar reactions of oxygen or nitrogen acids having comparable pK_a values (Walker et al., 1983). Presumably, substantial structural changes of the metal complexes, accompanying the deprotonation, are responsible for this sluggishness (Jordan and Norton, 1982). For example, the initially octahedral d^6 $Os(CO)_4LH$ complexes ($L = H$ or CH_3) become d^8 $[Os(CO)_4L]^-$ and approximately trigonal bipyramidal upon deprotonation. Such restructuration is caused by considerable electronic changes: the negative charge on the complex anion is delocalized into the π-acceptor ligands, as it is reflected in a remarkable decrease (> 100 cm^{-1} in some cases) in the carbonyl stretching frequencies for the anions, compared to the corresponding neutral hydrides. In contrast, deprotonation of oxygen or nitrogen acids generate a negative charge highly localized on these more electronegative ions.

Another facet of the problem of the M−H bond polarity is elucidated by the work of Messmer et al. (1977), who calculated the electronic structure of small Ni, Pd, and Pt clusters containing hydrogen (self-consistent-field X-alpha scattered wave method). The orbital energies resulting from these calculations are shown in Fig. 2.5, for the transition metal cluster with and without interstitial hydrogen. In the Pd case, the $1s$ hydrogen atomic orbital is located in the middle of the d-orbital range; in the Ni case it lies below, and in the Pt case at the upper end of the cluster d-orbitals. One consequence thereof is that the hybrid metal orbital responsible for the M−H bond is mainly made up of the $4s$ orbital in the Ni case, with some $3d$ and only minor $4p$ contribution, whereas the corresponding metal hybrids are mainly $4d$ and $5d$ in Pd and Pt respectively, as mentioned in Sect. 2.3.

Since the relative positions of the orbital energies of the hydrogen $1s$ and the bonding metal hybrid orbital are a measure of the differences in orbital electronegativity, these calculations indicate a small negative charge on the

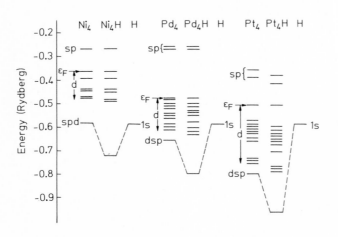

Fig. 2.5. SCF X α orbital energies of tetrahedral group VIII transition metal clusters with and without interstitial atomic hydrogen. ε_F = Fermi level (Messmer et al., 1977). Reproduced by permission of North-Holland Physics Publishing

hydrogen when bonded to Ni (in aggregate state), and a small positive charge on the hydrogen in the Pd and Pt cases. Taking into account, however, the near correspondence of the relevant orbital energies, one may anticipate that changes in the ligand environment of the metals may influence, or even reverse, the polarity of the bond. A most interesting illustration of this phenomenon was reported by Christmann and Ertl (1976): whereas H atoms on a flat Pt surface carry a small positive charge (in agreement with the calculations mentioned above), hydrogen atoms located in the vicinity of steps (cf. Fig. 2.2) are negatively charged:

$$Pt^{\delta-} - H^{\delta+} \qquad Pt^{\delta+} - H^{\delta-} \,.$$
(111) \qquad\qquad step

Another example of ligand influence on the M−H bond polarity may be seen in Table 2.2, Nrs. 10 and 11. The replacement of a CO ligand by $P(OCH_3)_3$, in otherwise structurally comparable complexes, reduces the acidity, decreasing the dissociation equilibrium constant K_a by three orders of magnitude. The phosphite ligand is a poorer π acceptor and a better σ donor than CO. Hence, the possibility of charge delocalization from the metal into the ligands is reduced in complex Nr. 11 (phosphite ligand), compared to complex Nr. 10 (CO ligand), resulting in less dissociation of a proton in the former case.

2.5 Mobility of Hydrogen Ligands

The migration of hydride ligands in metal complexes and clusters is well documented (Ugo, 1975; Band and Muetterties, 1978, and references therein). In most cases ^{1}H-NMR has been used to investigate fluxional behavior of hydride ligands in such compounds. For example, the trinuclear complex $\{HRh[P(OCH_3)_3]_2\}_3$ (I) has a hydride resonance that consists of a septet (spin-spin coupling with all six phosphorus atoms) of quartets (coupling with all three rhodium atoms), showing that the hydride ligands fluctuate very rapidly, on the NMR time scale. The hydride resonance is invariant down to − 90 °C.

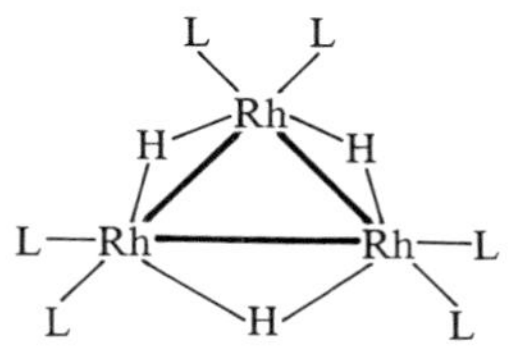

L = P(OCH₃)₃

I

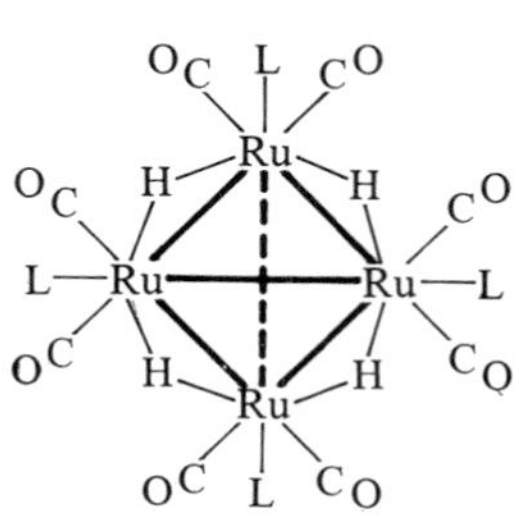

L = P(OCH₃)₃ or CO

II

Another example is provided by the series of tetrahedral compounds of formula

$$H_4Ru_4(CO)_{12-n}L_n \; [II, \, n = 1-4, \, L = P(OCH_3)_3]$$

which only show one hydride proton resonance from $+25$ to $-100\,^{\circ}C$, regardless of the degree of substitution. Thus, there is a rapid motion of the hydrides. The activation energy of these processes is low, $12-20$ kJ/mol, showing clearly that no complete bond breaking is involved. Exchange from bridge-bonded to end-on bonded positions and vice versa is one possible pathway; threefold bonded positions (centered over the Rh_3 triangle in I, or over the tetrahedron faces in II above) may also be involved. A synchronous motion of all mobile hydride ligands is generally assumed, and has been demonstrated in the case of a chiral Ru_4 cluster (Shapley et al., 1977).

In certain selected cases, EPR permits the observation of fluxional behavior of hydride ligands (Henrici-Olivé and Olivé, 1969). The EPR signal of the Ti/Al complex

$$\begin{array}{ccc} Cp & \diagdown \; H \; \diagup & \diagup Cl \\ & Ti \;\; Al & \\ Cp & \diagup \; H \; \diagdown & \diagdown Cl \end{array} \qquad Cp = \eta^5\text{-}C_5H_5$$

in THF consists of a sextet (interaction of the unpaired electron of Ti(III) with the nucleus of Al, $I = 5/2$) of triplets (interaction with two equivalent H); each of the 18 lines is further split into a multiplet, showing interaction with the 10 aromatic hydrogen nuclei of the two cyclopentadienyl rings. On cooling the sample, an alternating line-width effect involving the $M_I = \pm 1$ and $M_I = 0$ spin states of the two "equivalent" bridge hydrogen atoms is observed (central group of lines of each triplet broadened, first and third groups broadened, see Fig. 2.6). This indicates the existence of a dynamic molecular process in which the involved coupling constant undergoes an out-of-phase modulation. In view of the hydride fluctuation described above, an exchange of the two hydride ligands, via end-on and bridged positions, is the most probable interpretation of the phenomenon, expecially because a similar activation energy ($\simeq 16$ kJ/mol) has been estimated.

The hydrogen mobility on metal surfaces is also well documented, and is demonstrated most convincingly by the so-called spillover effect. The yellow tungsten oxide WO_3, for instance, does not react with hydrogen at ambient temperature and pressure. If, however, Pt is supported on WO_3, the "initiating phase" (in this case Pt) dissociates the hydrogen producing hydrogen ligands which then migrate to the "acceptor phase" (in this case the WO_3 support), and the dark blue hydrogen tungsten bronze H_xWO_3, is formed. In samples containing only 1% Pt on WO_3, sufficient hydrogen is transferred (spilt over)

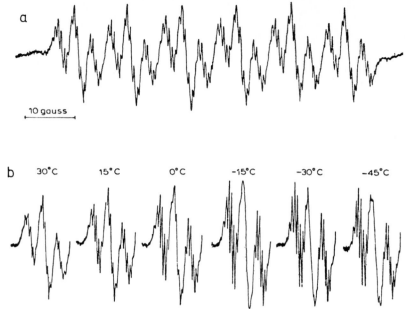

Fig. 2.6. EPR signal of $(C_5H_5)_2TiH_2AlCl_2$ in THF; a) entire spectrum at room temperature; b) alternating line width effect at lower temperature (first low field triplet of the spectrum) (Henrici-Olivé and Olivé, 1969). Reproduced by permission of Verlag Chemie

to the support to lead to an average composition of $H_{0.4}WO_3$, within 10 min (Sermon and Bond, 1980). No information is available on the actual size of the platinum crystallites on the support, and therefore of the number of "initiating" sites per WO_3 particle; so it is not possible to estimate an average distance travelled by the spilt-over hydrogen.

Palladium has a comparable spillover effect, whereas Au and Cu are less active and require temperatures of $> 200\,°C$. This reflects the relative activity of these metals in dissociating H_2.

In special cases where suitable paramagnetic species (notably Ti^{3+}) develop in the acceptor phase, EPR can be used as a tool to study hydrogen mobility. Thus the reversible formation of Ti^{3+} has been observed in the systems M/TiO_2 (M = Rh, Pt, Pd), at room temperature (Conesa and Soria, 1982; De Canio et al., 1983).

The described phenomena, whereby "initiating" and "accepting" phases are in close contact, is called *"primary spillover"*. Interestingly, a certain transfer of hydrogen is possible even if these phases do not have extended areas of contact, as for instance in mechanical mixtures of Pt/Al_2O_3 particles with WO_3 particles (in this system the spillover effect was in fact first observed, by Khoobiar, 1964). The reaction rate of this *"secondary spillover"* is much lower, but increases if the particle size is decreased by grinding (Sermon and Bond, 1980).

The actual mechanism of hydrogen mobility on metal surfaces is an interesting and not yet elucidated question. The situation is certainly somewhat different from the symmetric cluster compounds mentioned above, where the concerted simultaneous movement of all involved H ligands, with gradual transition from electron deficient bridge-bonds (half bonds) to full single bonds and vice versa does not change the valency state of the individual metal atoms. In surface mobility, however, redox reactions are involved. Thus, in hydrogen spillover, hexavalent tungsten is reduced to a valency state of $6-x$ in H_xWO_3, and Ti^{4+} is reduced to Ti^{3+}. Moreover, the "initiating phase" is transformed to its original metallic state in the process, since repeated subsequent dissociative additions of hydrogen appear to have taken place at each active Pt atom in the Pt/WO_3 case. Presumably, the hydrogen "travels" as an atom and ends up in the "acceptor phase" as a proton, donating its electron to the acceptor metal. On a plain metal surface, on the other hand, the process may be more similar to that in the cluster compounds, i.e. changes from single bonded to bridge bonded situations:

$$\underset{M(O)-\underset{\mid}{M}(I)}{\overset{\overset{H}{\mid}}{}} \rightleftarrows \underset{M - M}{\overset{.^{.}H._{.}}{}} \rightleftarrows \underset{\underset{\mid}{M}(I) - M(O)}{\overset{\overset{H}{\mid}}{}}.$$

2.6 Conclusion

Summarizing Chapter 2, we conclude that hydrogen bonded to a transition metal surface is in all important aspects similar to hydrogen in transition metal hydrido-complexes.

2.7 References

Alemdaroglu NH, Penninger JML, Oltay E (1976) Monatsh Chem 107:1043
Band E, Muetterties EL (1978) Chem Rev 78:639
Bau R, Teller RG, Kirtley SW, Koetzle TF (1979) Acc Chem Res 12:176
Bau R, Ho DM, Gibbins SG (1981) J Amer Chem Soc 103:4960
Bond GC (1966) Discuss Faraday Soc 200
Chen HW, Jolly WL, Kopf J, Lee TH (1979) J Amer Chem Soc 101:2607
Christmann K, Ertl G (1976) Surface Sci 60:365
Christmann K, Ertl G, Pignet T (1976) Surface Sci 54:365
Christmann K, Behm RJ, Ertl G, van Hove MA, Weinberg WH (1979) J Chem Phys 70:4168
Conesa JC, Soria J (1982) J Phys Chem 86:1392
DeCanio SJ, Apple TM, Dybowski CR (1983) J Phys Chem 87:194
Demuth JE (1977) quoted in Messmer et al. (1977)
De Vries B (1962) J Catal 1:489
Döbereiner JW (1836) Zur Chemie des Platins. P. Balz'sche Buchhandlung, Stuttgart; as
    quoted by Vaska and Werneke (1971)

Edgell WF, Gallup G (1956) J Amer Chem Soc 78:4188
Fassaert DJM, van der Avoird A (1976) Surface Sci 55:291
Fassaert DJM, Verbeck H, van der Avoird A (1972) Surface Sci 29:501
Henrici-Olivé G, Olivé S (1969) J Organometal Chem 19:309
Henrici-Olivé G, Olivé S (1976) Topics in Current Chemistry 67:107, and references therein
Henrici-Olivé G, Olivé S (1977) Coordination and Catalysis. Verlag Chemie, Weinheim, New York
Ibbotson DE, Wittrig TS, Weinberg WH (1980) J Chem Phys 72:4885
Jordan RF, Norton JR (1982) J Amer Chem Soc 104:1255
Khoobiar SJ (1964) J Phys Chem 68:411
Martin AJ (1978) Surface Sci 74:479
Messmer RP, Salahub DR, Johnson KH, Yang CY (1977) Chem Phys Letters 51:84
Pearson RG, Ford PC (1982) Comments Inorg Chem 1:279
Schmidt R, Geis M, Kelm H (1974) Z physik Chem, Neue Folge 92:223
Sermon PA, Bond GC (1980) J Chem Soc Faraday I 76:889
Shapley JR, Richter SI, Churchill MR, Lashewycz RA (1977) J Amer Chem Soc 99:7384
Shustorovich E, Baetzold RC, Muetterties EL (1983) J Phys Chem 87:1100
Ugo R (1975) Catal Rev Sci Eng 11:225
Vaska L, Werneke MF (1971) Trans NY Acad Sci 31:70
Walker HW, Pearson RG, Ford PC (1983) J Amer Chem Soc 105, 1179

3 Transition Metal-Carbon Monoxide Interactions

3.1 The CO Molecule

In contrast with hydrogen, carbon monoxide interacts with metals predominantly as an intact molecule. Therefore, some of the properties of the isolated CO molecule, in particular its electronic structure, shall be discussed in this section.

Carbon and oxygen bring the same set of atomic orbitals into the molecular orbitals. However, due to the greater effective nuclear charge of the oxygen atom, its atomic $2s$ and $2p$ orbitals are lower in energy than those of the corresponding carbon orbitals. In fact, the oxygen $2p_z$ level (z = direction of bonding) gives strong bonding overlap with the carbon $2s$ orbital. Thus, the oxygen $2p_z$ and the carbon $2s$ orbitals are more involved in the σ-bonding between the two atoms than are the oxygen $2s$ and carbon $2p_z$. On the other hand, the lone pair orbital at oxygen has more $2s$, that on carbon more $2p_z$ character. The bonding σ, as well as the degenerate pair of bonding π orbitals are heavily weighted on the oxygen side, the unoccupied σ^* and π^* orbitals are weighted correspondingly on the carbon atom (see Fig. 3.1). The energy levels and orbital population for the isolated CO molecule, as calculated by Bullett and O'Reilly (1979) are given in Table 3.1. Experimental gas phase

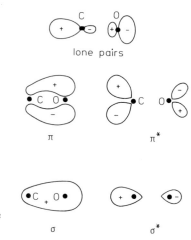

Fig. 3.1. Approximate boundary surface pictures of the MO's of CO (Henrici-Olivé and Olivé, 1977, p. 92)

Table 3.1. Energy levels and orbital populations for the isolated CO molecule according to Bullett and O'Reilly (1979).

Orbital	Energy (eV)	% on C	% on O	Transition state energy (eV)	Experimental Ionization Potential (eV)
5σ	-10.9	82	18	-14.8	14.0
1π	-12.6	23	77	-17.9	16.8
4σ	-15.6	26	74	-20.5	19.7
3σ	-27.7	9	91	-33.2	38.0

ionization potentials are included in the Table for comparison. The difference between the latter and the calculated orbital energies is due to the fact that a "transition state model" in which the appropriate molecular orbital contains only half an electron is required for estimating ionization potentials. The corresponding calculated transition state energies in Table 3.1 lie close to the ionization potentials. The highest occupied level is the 5σ state which is predominantly localized at the carbon end of the molecule and protrudes from it in the direction opposite to the C−O bond. It roughly corresponds to the carbon lone pair.

Note that the approximate boundary surfaces of the lone pairs and σ orbitals depicted in Fig. 3.1 do not correspond exactly to the molecular orbitals in Table 3.1, but the former are rather approximate representations of symmetry permitted linear combinations of the latter. (This does not apply to the π orbitals, where no symmetry permitted linear combination is available.) Such "manipulation" of wave functions may seem somewhat confusing to an experimental chemist. However, in a very rough manner one may visualize the situation as follows: the one and definite "electron cloud" of a given molecule can be formally divided into different regions by mathematical procedures, so that different calculation methods may give different patterns. For the treatment of certain molecular or chemical properties of the molecule, the one or the other description might then be more advantageous (see e.g., Henrici-Olivé and Olivé, 1977, Chap. 6.2).

The electron population in the orbitals (cf. Table 3.1 and Fig. 3.1) clearly indicates that the atomic charges have a C^+O^- polarity, in full accord with electronegativity considerations. On the other hand, the CO molecule has a very low dipole moment of 0.112 D, and the dipole polarity is C^-O^+, opposite to the charge polarity. This apparent inconsistency has been discussed by Politzer et al. (1981), emphasizing that a dipole moment does not necessarily reflect an actual charge distribution. An important reason for this is that the dipole moment contains distances as well as charge factors. The carbon lone pair makes a very significant contribution to the dipole moment, which is opposite to the contribution of the bonding electrons. The average position of the carbon lone pair electrons is far outside of the carbon, approximately 26 pm. (The bond length of CO is 112.8 pm.) Thus, the dipole moment is so low and the dipole polarity is C^-O^+ due to a fortuitous cancellation of opposing electronic contributions.

Atomic charges, on the other hand are designed to reflect the total quantity of electronic charge associated with each nucleus. On this basis, of course, the six bonding electrons, with an average position in the internuclear region, about 3/4 of the way from the carbon to the oxygen, are the dominant factor leading to C^+O^-.

However, Politzer et al. also pointed out that calculated electronic charges, or physically observable electronic density functions, may be misleading with regard to reactive sites and regions in a molecule, since they take no explicit account of the effects of the nuclei. A possible reaction partner is affected by a total charge distribution, not just by the electronic part of it. The authors calculated the electrostatic potential produced by its electrons and nuclei in the space around the CO molecule. An approximate sketch of their findings is given in Fig. 3.2. Two regions of negative potential located at the ends of the molecule can be seen. Though the most negative potential is found near the oxygen, both ends of the CO molecule have some tendency to interact with electrophilic species. This is a very interesting conclusion which evidently could not have been reached from atomic charge considerations, since one of the atoms must necessarily be found to have a positive charge.

Regarding the most probable site of reaction, however, one has to bear in mind that the carbon lone pair is the highest occupied orbital, i.e. it houses the least tightly bound electrons which should be available most readily for any shifting of charge towards a reaction partner; moreover, the carbon lone pair protrudes further into space than the oxygen lone pair (Fig. 3.1), making the overlap with a suitable electron acceptor orbital of a reaction partner more favorable.

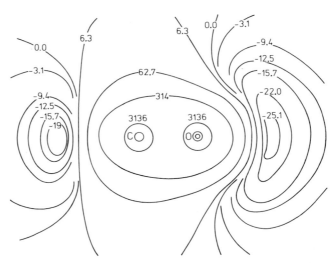

Fig. 3.2. Approximate sketch of the electrostatic potential of carbon monoxide in the plane passing through the molecular axis, as calculated by Politzer et al. (1981). Values are in kcal/mol (1 kcal = 4.187 kJ). Reproduced by permission of the American Chemical Society

3.2 Coordination Chemistry of CO in Molecular Complexes and Clusters

The first metal carbonyl compound, $Ni(CO)_4$, was discovered by Mond et al. in 1890. Since that time the chemistry of metal carbonyls has developed into one of the most intensively studied fields of inorganic and organo-metallic chemistry. An enormous number of novel and unusual structures are discovered every year, ranging from simple binary carbonyls (containing only one zerovalent metal and CO ligands), over cluster compounds made up of up to 19 metal atoms and over 30 CO ligands, derivatives of metal carbonyls containing other than CO ligands, to ionic carbonyls and clusters where the metal bears a formal charge and a gegenion is present. The formation, structure and chemistry of metal carbonyls, as well as their use in homogeneous catalysis, has been amply reviewed in many excellent articles (see e.g. Basolo and Pearson, 1967; Calderazzo et al., 1968; Cotton, 1976; Wender and Pino, 1977; Colton and McCormick, 1980; Benfield and Johnson, 1981). *In this Section, we shall focus on those electronic, structural, spectroscopic and thermodynamic details that will be of interest in the discussion of the catalytic CO hydrogenation.*

By far the prevailing bonding mode between a CO molecule and a transition metal atom or ion is with the carbon bonded to the metal, as born out by innumerable X-ray single crystal structural analyses. As a neutral molecule, CO does not establish a normal covalent bond with a transition metal center, where each of the two partners would contribute one electron. The bonding is rather of the "coordinative" type, which plays an important role in transition metal chemistry. It is characterized by the fact that both electrons come from one partner. A bonding and an antibonding molecular orbital (MO) arise from the interaction of an occupied orbital of one and an unoccupied orbital of the other partner. The bonding MO takes up the two electrons, the antibonding remains empty. Because the bonding MO is lower in energy than any of the constituent orbitals, the bonded situation is energetically more favorable than that with the two electrons in the isolated orbital of the donor partner.

In the particular case of metal−CO bonding, the donor orbital is the lone pair orbital at the carbon. The electronic structure of the CO molecule (Fig. 3.1) shows intuitively why this is so. The carbon lone pair not only protrudes further into space and hence is sterically better available for bond formation, it also is the highest occupied CO orbital, and any charge transfer towards the metal can be imagined to take place most readily with the least tightly bound electrons.

The carbon lone pair and a suitable empty d orbital of the metal (e.g. $d_{x^2-y^2}$ in an octahedral complex) have σ symmetry with regard to the bonding axis metal-ligand (no change of sign on rotation around this axis), and a σ bond results (see Fig. 3.3). Since the carbon provides the two electrons making up this bond, there is a certain electronic flow from the CO ligand to the metal. But there are also low lying unoccupied orbitals available at the CO molecule, namely the degenerate pair π^*. One of these orbitals has the right symmetry properties to match with the d_{xy} orbital (see Fig. 3.3), and to form a π MO

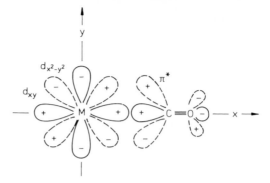

Fig. 3.3. Metal-CO bonding: σ bond between the carbon lone pair and a metal $d\sigma$ orbital and π back donation from a metal $d\pi$ orbital into the antibonding π^* at CO (d orbital notation for an octahedral complex)

(change of sign on rotating by 180° around the bond axis). The longer lobes at C are particularly suited for this purpose. The other antibonding π^* orbital of the CO molecule matches in the same way with d_{xz}. Hence a CO ligand may in principle form a coordinative σ bond and one or two coordinative π bonds with a transition metal, provided $d_{x^2-y^2}$ is empty and d_{xy} and/or d_{xz} are occupied. In the σ bond electron density is donated from the ligand to the metal, in the π bond vice versa ("electron back donation").

Certain metal centers can form such coordinative bonds with up to six carbon monoxide ligands, arranged in an octahedron, whereby the d_{z^2} and $d_{x^2-y^2}$ metal orbitals pointing to the apices of the octahedron are involved in the σ bonding, and d_{xy}, d_{xz} and d_{yz} in π bonding. Examples are $Cr(CO)_6$ and $W(CO)_6$. It is easily remembered (although not as simply visualized) that in tetrahedral complexes such as $Ni(CO)_4$ it is vice versa; d_{xy}, d_{xz}, d_{yz} are σ orbitals, and $d_{x^2-y^2}$, d_{z^2} participate in the π bonding to the ligands (see e.g. Orchin and Jaffe, 1971). Of course, not only the nd, but also the $(n+1)\,s$ and p orbitals of the metal are involved in either bonding pattern (see e.g., Beach and Gray, 1968).

The CO molecule as a ligand is said to be a typical π acceptor, meaning that the π interaction is much more important for the M–CO bond than the σ interaction. More charge is transferred in the π bond from the metal to the ligand than in the σ bond from the ligand to the metal. ESCA data, as well as theoretical calculations, indicate a formal positive charge on the metal in neutral complexes such as $Cr(CO)_6$, $Fe(CO)_5$ and $Ni(CO)_4$ (Baerends and Ros, 1975).

An important consequence of the presence of electrons in the antibonding π^* orbital of CO is the reduction of the C–O bond strength which can, to a first approximation, be estimated from the CO stretching frequency. Depending on the symmetry of a particular carbonyl complex, the CO stretching vibration gives rise to one or several strong sharp absorption bands in IR. The stretching frequency v_{CO} of CO itself has a value of 2155 cm^{-1}. In carbonyl complexes, downward shifts by up to several hundred cm^{-1} are not unusual, see Table 3.2 where some representative examples are summarized. For the neutral carbonyl complexes, v_{CO} has values in the neighborhood of 2000 cm^{-1}.

Table 3.2. Stretching frequency v_{CO} of terminal CO ligands for mononuclear carbonyl complexes (Basolo and Pearson, 1967).

Compound	v_{CO} (cm^{-1})[a]			
$Cr(CO)_6$	$\simeq 2100$,	$\simeq 2000$,	<u>1981</u>	
$Fe(CO)_5$	2034,	2014		
$Ni(CO)_4$	2057			
$[Co(CO)_4]^-$	1886			
$[Fe(CO)_4]^{2-}$	1786			
$Mo(CO)_6$	2120,	2022,	<u>1990</u>	
$[Mo(CO)_5P(CH_3)_3]$	2071,	1952,	1943	
cis $Mo(CO)_4[P(CH_3)_3]_2$	2019,	1920,	1903,	1893

[a] For the two symmetric octahedral complexes, the very strong "allowed" (T_{1u}) vibration is underlined.

A small but significant increase in the C−O bond length (e.g. 116 pm and 115 pm for CO in $Cr(CO)_6$ and in $Ni(CO)_4$, respectively, versus 113 pm for free CO, Hillier and Saunders, 1971) accompanies the decrease of v_{CO}. Every event that enhances electron back donation into the π^* orbital reduces v_{CO} further. This is particularly evident if the oxidation state of the metal is changed. In the isoelectronic series $Ni(CO)_4$, $[Co(CO)_4]^-$, $[Fe(CO)_4]^{2-}$, the negative charge is increasingly delocalized into the CO ligands, decreasing the C−O bond order. But also the substitution of one or several CO ligands by other ligands that are poorer π acceptor than CO reduces the bond order, as exemplified by the last three complexes in Table 3.2. Evidently, the remaining CO ligands must accept $d\pi$ electrons density to a greater extent if the σ donor ligand $P(CH_3)_3$ replaces one or two CO ligands.

Vibrational spectroscopy in the range of the v_{CO} stretching frequency is in fact a well established routine tool in the investigation of metal carbonyl complexes. The bands are generally sharp and very intense; as an example, the intensity of the allowed T_{1u} vibration of the octahedral $Cr(CO)_6$ complex at 1981 cm^{-1} (see Table 3.2) is 100 times that of the free CO (Brown and Darensbourg, 1967). Moreover, the CO vibrational bands are well separated from other vibrational modes of the complexes: the M−C stretches and M−C−O bends which occur around 650−350 cm^{-1}, and the C−M−C bends which are found below 150 cm^{-1}. As shown by the few examples in Table 3.2, v_{CO} is highly sensitive to symmetry, formal charge, and the influence of other ligands.

Force constants* derived from v_{CO} have often been linked to the CO bond order in a quantitative way by associating the force constant of carbon monoxide with a bond order of three, that of an organic carbonyl (such as formaldehyde) with a bond order of two, and finding that of CO in a metal

---

* The force constant, f, of a bond between atoms A and B is given by $f = v_{AB}\,\mu/17.0$, where v_{AB} is the vibration frequency of the bond in cm^{-1} and μ is the reduced mass of the A-B oscillator in atomic weight units.

carbonyl by interpolation. But Braterman (1976) has cautioned that this procedure may be quite misleading. Thus, a bond order of ca. 2.5 was found for CO in $V(CO)_6^-$ by this way, which is in conflict with all expectation. It would mean that the six valence electrons of V^- should be assigned entirely to the six CO ligands, so that the compound would be a complex of vanadium in the $+5$ oxidation state with six CO^- ligands. At the origin of such misinterpretations appear to be mainly two facts. On the one side, the limiting value for bond order two may be unreliable. This is indicated by the fact that the CO force constants for formaldehyde, acetaldehyde and acetone are different and show that force constants are sensitive to changes that do not affect π bond orders. On the other hand, "orbital following" may play an important role. If the nuclei are displaced in a vibrational movement, out of their equilibrium position, the "electron cloud" will rearrange concomitantly in order to relax to the lowest energy distribution in the field of the displaced nuclei. This process of orbital following will lower the energy of the distorted molecule, and hence lower the force constant. Apparently, CO ligands in carbonyl complexes exhibit a degree of orbital following considerably more marked than in free CO or organic carbonyl compounds. The force constants are then much lower than any correlation with bond order would require. Incidentally, the strong orbital following is also responsible for the high intensity of the CO stretching bands in carbonyl complexes. As a CO group is stretched, the energy of the CO π^* orbital is lowered and back donation increases, giving rise to a large oscillating dipole.

Though not for the determination of absolute bond orders, the CO stretching frequency is yet a very important tool in the investigation of binary and substituted carbonyl complexes, providing information as to *cis* and *trans* effects of different ligands, the relative importance of electron density in σ and π states for these effects, as well as to the presence of geometric isomers (Braterman, 1976).

One of the most fascinating facets of transition metal carbonyl chemistry is the existence of a very large number of bi- and polynuclear homo- and heteronuclear metal carbonyl complexes, or transition metal cluster carbonyls (see e.g. Johnson, 1980). The expression *cluster carbonyls* is generally used for those transition metal carbonyl complexes having two or more metal atoms, and metal-metal bonds between them. X-ray crystal structures are known for many of them, and have revealed that these clusters consist of a metal core in which the constituent metal atoms are linked together by metal–metal bonds. The metal core for clusters with three and more metal atoms is in general highly symmetric, with a triangle for three, a tetrahedron for four, a trigonal bipyramide for five, and an octahedron for six metal atoms being the most frequently found arrays. The carbonyl ligands surround this core, whereby CO has a variety of bonding possibilities (see Fig. 3.4). Apart from the terminal bonding as in $Mn_2(CO)_{10}$ and $Os_3(CO)_{12}$, there is bridge bonding between two metal centers, as in $Fe_2(\mu\text{-}CO)_3(CO)_6$ and triple bridge bonding (face bridging) as in $Cp_4Fe_4(\mu_3\text{-}CO)_4$. Still a different bonding type has been reported by Commons and Hoskins (1975). When $Mn_2(CO)_{10}$ was refluxed in n-decane with $Ph_2PCH_2PPh_2$ (dpm), a bidentate ligand, the complex

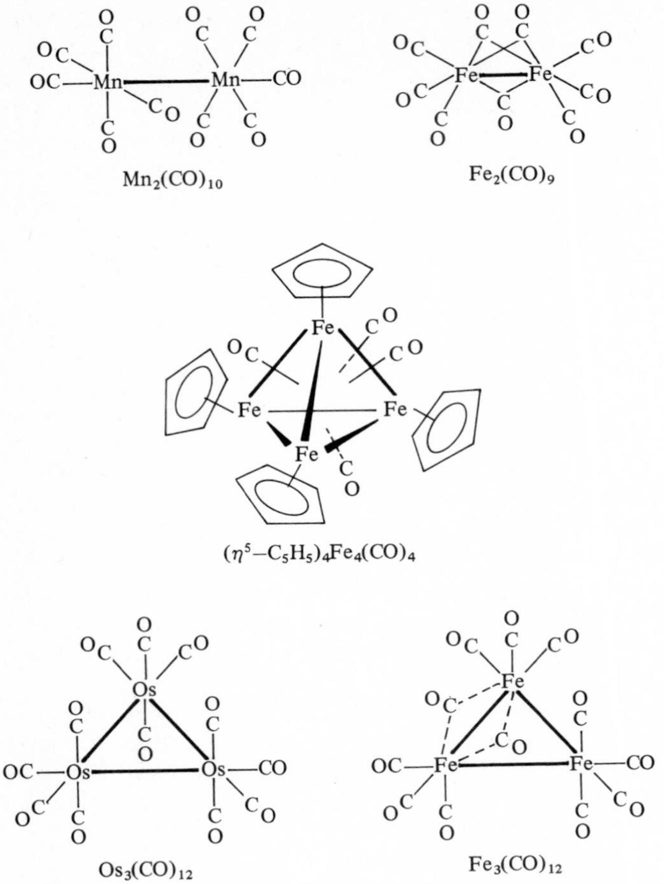

$Mn_2(CO)_{10}$

$Fe_2(CO)_9$

$(\eta^5-C_5H_5)_4Fe_4(CO)_4$

$Os_3(CO)_{12}$

$Fe_3(CO)_{12}$

Fig. 3.4. Structures of some polynuclear transition metal carbonyls (see e.g., Cotton and Wilkinson, 1972)

$Mn_2(CO)_5(dpm)_2$ was formed. Together with four IR bands that could be assigned to terminal stretching vibrations, the IR spectrum of the complex showed a strong band at 1645 cm^{-1}. A single crystal X-ray diffraction analysis revealed that one of the CO ligands, σ-bonded to one of the Mn centers, is at the same time π-bonded (dihapto coordinated) to the other:

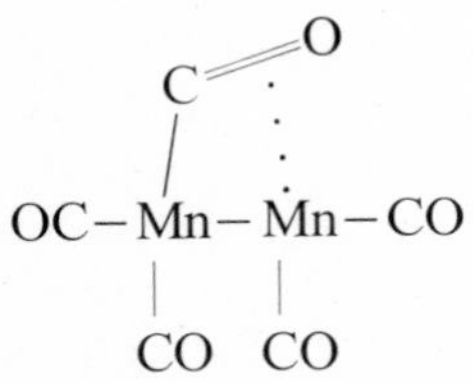

(each of the two bidentate dpm ligands is coordinated to both Mn centers). Amazingly, the C−O bond distance of this special CO ligand (110 pm) is not longer than that of the others in the complex. A dihapto coordinated CO ligand was also observed by Manassero et al. (1976). Protonation of the tetrahedral iron cluster [Fe$_4$(CO)$_{12}$(μ_3-CO)]$^{2-}$, containing a face-capping CO ligand, did not lead to the protonation of CO, but rather to a metal-bridging hydride ligand. But at the same time the iron core suffered structural rearrangement and the face bridging μ_3-CO became dihapto bonded to one of the iron centers:

$$\text{(3.1)}$$

The new structure of the iron core was named a "butterfly" complex by Manassero et al. In this case, the special CO ligand exhibits a relatively long C−O bond distance of 126 pm (typical cluster C−O distances are 110−114 pm for CO ligands bonded only through the C atom).

The rules governing the cluster size as well as the presence of terminal and bridging CO ligands are not completely elucidated. However, the arrangement of the CO ligands in highly symmetric polyhedral geometries, in the crystalline state, appears to be of major importance (Benfield and Johnson, 1981). Evidently, the metal cluster itself has to fit into the polyhedral interstice, hence the size of the metal centers may also play a role. Thus, it has been found that for Mn$_2$(CO)$_{10}$ the two-metal unit is enveloped by a bicapped square antiprism of CO ligands; the 9 CO ligands of Fe$_2$(CO)$_9$ form a tricapped trigonal prismatic polyhedron housing the Fe−Fe unit; for Os$_3$(CO)$_{12}$, the triangular Os cluster fits into an anticubo-octahedral array of the 12 linearly bonded CO ligands, whereas for Fe$_3$(CO)$_{12}$ the ligand arrangement is icosa- hedric, giving rise to a doubly bridged structure between two of the Fe centers (cf. Fig. 3.4). In general, the third row transition metals tend towards solely terminal CO coordination, whereas μ_2-bridged structures are more often found in the earlier metals, and μ_3-CO appears to be restricted to Fe and Ni carbonyl compounds.

Electronic influences may also contribute to the overall arrangement. A comparison of X-ray structures and IR spectra of the complexes has shown that terminal M−CO groups usually have CO stretching frequencies in the region 1900−2150 cm^{-1} (in neutral molecules, cf. Table 2.4), whereas bridging CO ligands fall in the range 1750−1850 cm^{-1}, and triply bridging CO ligands in the range 1620−1730 cm^{-1} (Chini et al., 1976), indicating increasing weakening of the C−O bond. Bridging CO ligands are, evidently, better π acceptors than terminal ones, hence may serve to delocalize surplus of negative charge from clusters which are anionic, or contain first row transition metals with low ionization potentials. More importantly, carbonyl complexes and clusters tend to obey the 18 electron ("noble gas") rule (Chini, 1970), i.e. the number of electrons in the valence shell of a constituent metal atom plus the lone pair electrons of the CO ligands attached to it (two per linearly bonded CO, one per bridge bonded) plus the number of metal-metal bonds starting from that particular metal atom sums up to 18.

In most cases, such considerations of the symmetry of the metal core and the carbonyl shell, as well as of the electronic situation permit to rationalize why a particular cluster structure is stable; they do not, however, allow to predict the structure to be expected in a particular synthesis. With the exception of finely divided nickel, the metals do not react easily with CO, at room temperature and normal pressure. Generally, metal carbonyl complexes and clusters are prepared by treating a metal salt, e.g. carbonate or halide, suspended in an organic solvent such as tetrahydrofuran, with carbon monoxide at $200-300$ atm and temperatures up to $300\,°C$ in the presence of a reducing agent. Presumably, a carbonyl cluster forming under such severe conditions is able to rearrange until the energetically and geometrically most stable configuration is attained.

The foregoing discussion has emphasized the character of the metal$-$CO bond as a double bond. One might expect a very strong bonding resulting from this interaction. However, the metal$-$CO bonds in carbonyl complexes and clusters (coordinative bonding) are relatively weak, if compared with covalent carbon$-$CO bonds, as, for instance in ketones with dissociation energies of the order of $330-340\,kJ/mol$ (Egger and Cocks, 1973). Some representative experimental bond dissociation energies for metal carbonyls $M_x(CO)_y$ are given in Table 3.3. With a few minor inversions, the trend is clear: the bond strength increases with increasing atomic weight of the transition metal, and appears to be independent of the cluster size. Substitution of one or more CO ligands by other ligands, which are usually poorer π acceptors and better σ donors than CO, strengthens the remaining M$-$C bonds because under these circumstances there is less competition for π back donation (stronger M$-$C π bonds). Moreover, a CO ligand trans to a substituent is more strongly bonded than one trans to another CO ligand, because two trans CO ligands necessarily

Table 3.3. Experimental metal$-$CO bond dissociation energies $\bar{D}(M-CO)$ for gaseous metal carbonyls (Connor, 1977; Behrens, 1978).

Metal Carbonyl[a] $M_x(CO)_y$	$\bar{D}(M-CO)$ kJ/mol CO	Metal Carbonyl $M_x(CO)_y$	$\bar{D}(M-CO)$ kJ/mol CO
$Cr(CO)_6$	107	$Mo(CO)_6$	152
• $Mn(CO)_5$	99	$Ru_3(CO)_{12}$	167
$Mn_2(CO)_{10}$	99	$Rh_2(CO)_8$	165
$Fe(CO)_5$	119	$W(CO)_6$	178
$Fe_3(CO)_{12}$	117	$Re_2(CO)_{10}$	187
• $Co(CO)_4$	136	$Os_3(CO)_{12}$	191
$Co_2(CO)_8$	136	$Ir_2(CO)_8$	190
$Ni(CO)_4$	147	$Ir_4(CO)_{12}$	190

[a] • indicates a radical

Table 3.4. Structural data for some $Cr(CO)_nX_{6-n}$ complexes (Smith, 1976).

Compound	Cr−C distance (pm)	
	trans to CO	trans to X
$Cr(CO)_6$	190.9	−
$Cr(CO)_5PPh_3$	188.0	188.4
$Cr(CO)_5P(OPh)_3$	189.6	186.1
$Cr(CO)_4(P(OPh)_3)_2$	187.8	−
$Cr(CO)_3(PH_3)_3$	−	183.8

share the same metal d orbital for π bonding. Structural studies on a series of substituted Cr carbonyl complexes illustrate this situation (Table 3.4).

The bond dissociation energies in Table 3.3 are averages over all M−CO bonds present in a particular cluster compound. They are determined from the total enthalpy of disruption of the entire cluster, ΔH_D. In particular, they do not distinguish between linear and bridging CO ligands. Connor (1977) has estimated the dissociation energy of the two types of bonding in the following way. Assuming that the contributions to ΔH_D of each metal-linear CO, metal-bridged CO and metal-metal [$D(M-CO_1)$, $D(M-CO_b)$, and $D(M-M)$, respectively] are the same for all binary carbonyl clusters of the same metal, the knowledge of ΔH_D for three compounds suffices to determine the contributions separately. (Of course the crystal structure must be known also.) Thus, for the three iron carbonyls $Fe(CO)_5$ ($\Delta H_D = 585$ kJ/mol), $Fe_2(CO)_9$ ($\Delta H_D = 1173$ kJ/mol) and $Fe_3(CO)_{12}$ ($\Delta H_D = 1676$ kJ/mol) the solution of the three simultaneous equations

$$5D(Fe-CO_1) = 585$$
$$D(Fe-Fe) + 6D(Fe-CO_b) + 6D(Fe-CO_1) = 1173$$
$$3D(Fe-Fe) = 4D(Fe-CO_b) + 10D(Fe-CO_1) = 1676$$

led to $D(Fe-CO_1) = 117$ kJ/mol, $D(Fe-CO_b) = 64$ kJ/mol and $D(M-M) = 82$ kJ/mol. The corresponding values determined from three cobalt carbonyls are $D(Co-CO_1) = 136$ kJ/mol, $D(Co-CO_b) = 68$ kJ/mol and $D(Co-Co) = 83$ kJ/mol. The interesting result is that the strength of each of the two metal−carbon bonds of a bridging CO is roughly one half of that of a linearly bonded CO. Hence, there does not appear to be a great deal of energetic difference between the two bonding modes.

This estimate is intuitively consistent with the rapid scrambling of all CO ligands that has been found for many carbonyl complexes and clusters in solution, and assumed to take place via linear-bridge exchange. These effects can easily be studied by ^{13}C NMR since the equilibration takes place in the NMR time scale. In certain cases, a limiting spectrum is observed at low or very low temperatures, showing different signals for CO ligands in symmetrically different positions; at higher temperature then a full exchange situation is obtained. From such studies the activation energy of the exchange process can be determined, and has been found generally below 90 kJ/mol, in some cases even as low as 20 kJ/mol. Evidently, the equilibration occurs in a concerted way, in order that each metal center maintains a constant electron count throughout the process. Unsymmetric CO bridges may be involved (Cotton, 1976):

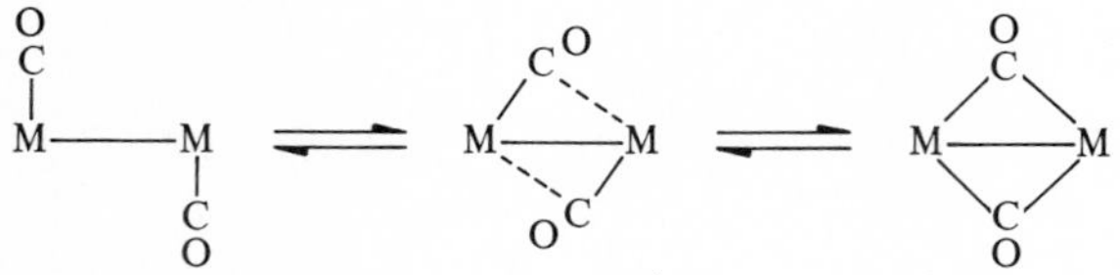

In certain complexes, such as for instance $Fe_3(CO)_{12}$ (see Fig. 3.4), the exchange between bridge and terminal positions is so facile that even at the lowest workable temperature there is only one single carbonyl resonance in the NMR spectrum. On the other hand, some polynuclear carbonyl clusters, in particular those of the higher transition metals, which are less prone to bridge formation, are stereochemically rigid up to 50 °C, as for instance $[Rh_{12}(CO)_{30}]^{2-}$ (Chini et al., 1974). The very symmetric complex $Os_3(CO)_{12}$ (see Fig. 3.4) shows two ^{13}C NMR signals of equal intensity at room temperature, which are assigned to the carbons in the axial and equatorial positions. At 150 °C, one single line is observed, and a local exchange of CO ligands at each Os atom was considered a possibility. However, it has been shown that all CO ligands have access to the three Os atoms, by using ^{187}Os (nuclear spin 1/2) and observing a 1:3:3:1 quadruplet in the ^{13}C spectrum at 150 °C (Koridze et al., 1981). We conclude that the CO scrambling in complexes and clusters of the lower transition metals is particularly favored as an interchange between different structures of almost equal energy. For most higher metals the energy difference between linear and bridge-bonding CO is still small enough for scrambling to take place at elevated temperature.

Although apparently similar from an energetic point of view, linear and bridging CO ligands are quite different in their reactivity towards Lewis acids. Shriver and coworkers (Alich et al., 1972) have shown that the complex $[(\eta^5\text{-}C_5H_5)Fe(CO)_2]_2$ which has two linear and two bridging CO ligands, forms a 2:1 adduct with $AlEt_3$:

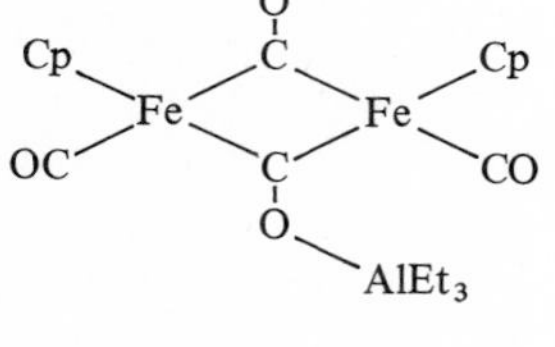

$Cp = \eta^5\text{-}C_5H_5$.

The two linear CO ligands do not react, even at large excess of the Lewis acid. Complexes with triply bridging CO such as $(\eta^5\text{-}C_5H_5)_4Fe_4(CO)_4$ (see Fig. 3.4), and the related $(\eta^5\text{-}C_5H_5)_3Ni_3(CO)_2$ with two triply bridging CO ligands above and below the Ni triangle, form 4:1 and 2:1 adducts respectively with aluminum alkyls. Not very many metal carbonyls are stable in the presence of aluminum alkyls; $Mn_2(CO)_{10}$ which is stable, but has no bridging carbonyls, does not react. The adduct formation of bridging CO ligands is accompanied by a considerable decrease of v_{CO} for the bridging CO (over $100\ \text{cm}^{-1}$) and, where applicable, by a concomitant increase of v_{CO} for the terminal CO, by ca. $40\ \text{cm}^{-1}$. Organic ketones form similar adducts accompanied by a decrease in v_{CO}, and one is tempted to compare the bonding of bridging CO with that occurring in the organic carbonyl compounds. However, the $M-C-M$ angle is generally much smaller than the sp^2 angle of $120°$. Braterman (1972) has argued against the idea of "ketonic" carbonyl ligands in favor of a delocalized molecular orbital description of the bonding which conceptionally appears to be better suited to interpret the large and opposite v_{CO} changes of linear and bridged CO ligands in the same complex, on adduct formation.

3.3 Molecular Carbon Monoxide on Metal Surfaces

Although the most straightforward and unambiguous method of investigating molecular metal carbonyl complexes, single crystal X-ray structural analysis, is not applicable to CO interacting with a metal surface, there are today a number of methods available that permit an atomic-scale scrutiny of surface monolayers (see e.g., Somorjai, 1981; White, 1983).

Infrared spectroscopy (IRS) is the most widely used tool, both because of its inherent simplicity and relatively low cost, and because of the often helpful comparison with the corresponding data for molecular complexes.

Whereas IRS is based on the vibrational excitation of surface atoms by absorption of infrared radiation, *high-resolution electron energy loss spectroscopy* (HREELS) uses inelastic reflexion of low-energy electrons for the same type of excitation. Similar information about structure and bonding of surface layers is obtained.

Auger electron spectroscopy (AES), monitoring the electron emission from the surface atoms excited by electron, X-ray, or ion bombardment, can yield an elemental analysis of a surface.

Ultraviolet and *X-ray photoelectron spectroscopy* (UPS, XPS) study the radiation stimulated emission from the valence shell of surface atoms, and can inform about the energy of electronic states of surface atoms and adsorbed species.

Angle resolved photoemission spectroscopy can be used to assess the orientation of molecules like CO on a "flat" metal surface.

Work function changes ($\Delta\Phi$) can inform about the direction of net charge transfer during bond formation (Nieuwenhuys, 1981), $-\Delta\Phi$ indicating electron flow from the metal to the adsorbed molecule (π-back donation in the case of adsorbed CO), $+ \Delta\Phi$ electron flow in the opposite direction. (The work function Φ of a metal, where e Φ is the work necessary to remove an electron from the highest populated level, can be obtained from photoelectron spectroscopy.)

Thermal desorption spectroscopy (TDS) investigates the thermally induced desorption or decomposition of adsorbed species, giving access to the energetics of adsorption, as well as to the composition of adsorbed species.

Finally, *low energy electron diffraction* (LEED), which measures the elastic back-scattering of low energy electrons by surface atoms, informs about symmetry features of the atomic surface structure and of adsorbed gases.

Although the list is by far not complete, these are the most commonly used methods up to date. Very often a combination of several methods is necessary in order to obtain a clear picture of a particular surface situation.

Surface science is concerned generally with the interaction of well-defined clean surfaces of single crystals with adsorbed species. Most of the methods of surface scientists require, apart from the highly sophisticated equipment, work under ultrahigh vacuum. Although these conditions are quite different from those found with "real" catalysts, a good part of the enormous host of knowledge gathered under these "aseptic" conditions can be carried over into the world of catalysis.

Adsorption of CO on group VIII metal surfaces normally takes place in molecular form, although sometimes a tendency to dissociation is observed, particularly for the iron group and at elevated temperature. For the higher group VIII metals, dissociation is considered spurious (Engel and Ertl, 1979). Dissociation and the formation of carbides will be discussed in the following Section. Here we concentrate on the interaction of the undissociated CO molecule with a transition metal surface. Surface scientists use the expression "associative adsorption" (as opposed to "dissociative adsorption") for this type of interaction. From the point of view of coordination chemistry it is better characterized by "coordinative bond formation", as it will be discussed below in some detail.

It was first suggested by Blyholder (1964), and is today commonly accepted that the interaction of a CO molecule with a transition metal surface leads to a bond formation quite similar to that in molecular carbonyls and clusters. From a conceptional point of view, this is in agreement with a more recent theory of transition metals (Pettifor, 1977; Woolley, 1980) that describes these metals in a simple and transparent molecular orbital framework. The energy bands formed from the valence *s* and *p* atomic orbitals are similar to those found in the non-transition metals; they are associated with quantum states in which the electrons move through the whole crystal almost like free particles in a box. The electron states formed from the valence *d* orbitals, on the other hand, are localized about the atomic centers, and are describable in the linear combination of atomic orbitals framework familiar in molecular quantum chemistry. Metal properties like cohesive energy per atom, compressibility and structural systematics across the Periodic Table are satisfactorily taken care of by this theory.

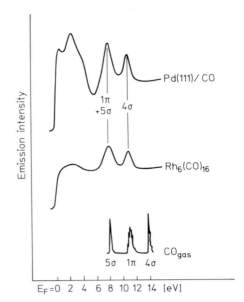

Fig. 3.5. Ultraviolet photoelectron spectra from a) CO on a Pd(111) surface; b) $Rh_6(CO)_{16}$; c) gaseous CO. Energy scale referred to the Fermi level E_F of the metallic system in a); onset of the d-band emission in b) lined up with that in a) (Conrad et al., 1976). Reproduced by permission of North-Holland Physics Publishing

We then can expect, in principle, the same type of bonding as described in Fig. 3.2: electron donation from the carbon lone pair into an empty metal d orbital of σ symmetry with regard to the metal−C bond, and electron back donation from a metal $d\pi$ orbital into the antibonding π* orbital of CO. A wide variety of experimental results supporting this view is available and has been reviewed by Engel and Ertl (1979). Thus the ultraviolet photoelectron spectra of CO on metal surfaces and of carbonyl complexes are almost identical (see Fig. 3.5). The accidental overlap of the CO 1 π and 5 σ orbitals in both cases is of particular interest: it signals the lowering of the energy of 5 σ (carbon lone pair) on bond formation with the metal, indicating a similar type of bonding. (The comparison of two different metals in Fig. 3.5 is admissible, because UPS of CO on different metal surfaces shows very little variation, as pointed out by Conrad et al., 1976.) Positive work function changes, $+ \Delta\Phi$, have been found for different crystallographic planes and/or films of the group VIII metals, indicating electron back donation from the metal into the antibonding π* orbital of CO (Nieuwenhuys, 1981, and references therein). Further information comes from angular resolved photo-electron spectroscopy, as well as from a LEED intensity analysis which have conclusively shown that CO is oriented perpendicularly on a number of different metal surfaces, with the carbon atom attached to the surface. Model calculations confirm that this arrangement is energetically the most favorable. Bond energies, as determined by TDS, are generally in the range of 100−170 kJ/mol, which is close to those shown in Table 3.3 for molecular complexes. Some examples of bond energies for CO on different crystal-lographic planes on metallic single crystals are shown in Table 3.5. The data indicate that the bond strength is quite independent of the geometry of the particular crystal face. For a given metal, it varies by less than 15%, and this holds for the most densely packed (111), (100), and (110) crystal planes as well as for the more open (210) and (311) planes. (The same is true for H_2 on metals, see Sect. 2.2.)

The data in Table 3.5 refer to very low coverage of the surfaces with CO. Presumably, they correspond to the energetically most favorable configurations on a particular plane. At higher coverage the CO molecules begin to interact, pushing one another into less favorable positions, and the bond energy goes down. As an example: on Pd(111) at low coverage Θ ($\Theta \leq 0.3$, i.e. less than one third of a monolayer) the symmetry features of LEED are best interpreted assuming the CO molecules located in threefold coordination sites. The

Table 3.5. Bond energy (in kJ/mol) for CO on single crystal planes (Engel and Ertl, 1979; Conrad et al., 1977).

Plane	(111)	(100)	(110)	(210)	(311)
Ni	111	–	125	–	–
Rh	130	121	–	–	–
Pd	142	153	167	148	146
Ir	142	146	155	–	–
Pt	138	134	109	–	–

adsorption energy (bond strength Pd−CO) is constant up to this coverage, ca. 142 kJ/mol. On increasing the coverage, the symmetry features change until at $\Theta = 0.5$ all CO molecules have moved into bridge positions; the bond energy has dropped by ca. 8 kJ/mol. Further coverage can be obtained only far below room temperature because the mutual repulsion between neighboring CO molecules becomes strong; this is accompanied by a steep decrease in the bond strength Pd−CO (Engel and Ertl, 1979).

The IR C−O stretching frequency, v_{CO}, for CO coordinatively bonded to metal surfaces also falls into the same ranges as in the cases of the molecular complexes and clusters. Actually v_{CO} is often used to confirm assumed bonding types on surfaces. Thus, for instance, the particular LEED features of the Pd(100) plane covered with CO up to $\Theta = 0.5$ could be explained only with the occupation of bridge sites. In fact IRS indicated the existance of only one species at very low coverage, with $v_{CO} \simeq 1900$ cm^{-1} which is in agreement with the assumed bridge bonding. However, the frequency increased continuously from $v = 1895$ cm^{-1} at very low coverage, to $v = 1949$ cm^{-1} at $\Theta = 0.5$. This was interpreted as due to the in-phase vibration of the whole coupled system of surface bonded CO molecules, because of intermolecular interactions (orbital overlap). Engel and Ertl (1979) therefore cautioned against the interpretation of IR spectra of surface bonded CO as characteristic for local bonding types, with the exception of very low coverages.

On the other hand, a linear end-on arrangement was reported by several authors for CO on a Ni(100) plane, at low coverage (Heinz et al., 1979 and

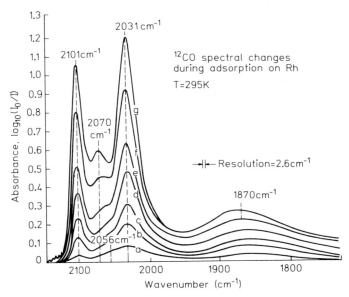

Fig. 3.6. Infrared spectra of CO on Rh/Al$_2$O$_3$, for increasing CO coverage; CO pressure (Torr): 2.9×10^{-3} (a), 4.3×10^{-3} (b), 5.0×10^{-3} (c), 8.3×10^{-3} (d), 0.76 (e), 9.4 (f), 50 (g); room temperature (Yates et al., 1979 a). Reproduced by permission of the American Institute of Physics

references therein). The stretching frequency, $v_{CO} = 2069$ cm^{-1} (Andersson, 1977) is very close to the corresponding frequency in Ni(CO)$_4$ (see Table 3.2). Also the bond lengths involved are almost identical, namely $d_{Ni-C} = 180$ pm and $d_{C-O} = 115$ pm for CO on Ni(100), and $d_{Ni-C} = 182$ pm and $d_{C-O} = 115$ pm for Ni(CO)$_4$ (Heinz et al., 1979).

Highly dispersed metals, on inert, high area support materials, evidently resemble more closely actual heterogeneous catalysts than do single crystal surfaces. Infrared spectroscopy appears to be to date the most suitable method to investigate the interactions of such systems with CO (Yates et al., 1979 a and b, and references therein). For this kind of experiments, a coating of support (e.g. Al$_2$O$_3$) and finely divided metal (e.g. Rh) is deposited onto the CaF$_2$ window of a vacuum-tight transmission IRS equipment. Inelastic electron tunneling spectroscopy (IETS) is also applicable to dispersed metals, giving information on low energy vibrational modes such as M−CO stretching and CO bending modes, which generally cannot be seen in IRS. Although not a routine method, IETS may be very useful in complementing IRS (Kroeker et al., 1980; Kroeker and Hansma, 1981).

The spectral changes observed by Yates et al. (1979 a) in the system Al$_2$O$_3$/Rh/CO, at varying CO coverage, are represented in Fig. 3.6. The most pronounced feature is a doublet with components at 2101 and 2031 cm^{-1}. In contrast to the other features, this doublet does not show any wave number shift with increasing CO coverage. The doublet was first reported by Yang and Garland (1957), and assigned to two CO molecules bonded to one single, isolated Rh

Table 3.6. Comparison of symmetric and antisymmetric carbonyl stretching frequencies for dicarbonyl rhodium species (Yates et al., 1979).

Species	CO Stretching frequency (cm^{-1})	
	v_{sym}	v_{asym}
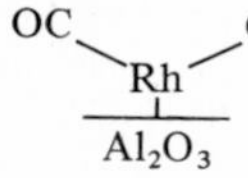 OC–Rh–CO on Al_2O_3	2101	2031
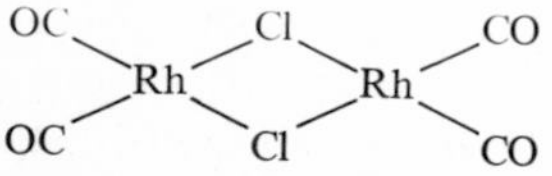 OC–Rh(Cl)(Cl)Rh–CO (OC, OC)	2095	2043

atom. The assignment was based on the close correspondance with a doublet observed in the spectrum of $Rh_2(CO)_4Cl_2$ which has two terminal CO ligands on each Rh center (see Table 3.6). This assignment is corroborated by the independence of the frequency on coverage: CO ligands on isolated metal centers are not expected to be influenced by CO molecules coordinated to vicinal metal centers. There is some evidence that the dicarbonyl rhodium species refers to Rh^+, as in the model complex in Table 3.6 (Worley et al., 1983; Yates and Kolasinski, 1983).

For molecular dicarbonyl complexes it has been found that the ratio of integrated absorbances for the antisymmetric and the symmetric component of the doublet is related to the angle, (2α), between the two carbonyl groups as follows:

$$A_{asym}/A_{sym} = \tan^2 \alpha.$$

For the doublet caused by the $Rh(CO)_2$ species on Al_2O_3, a value of $2\alpha = 90°$ results, in good agreement with the bond angle ($91°$) between the two carbonyl groups on each Rh in $Rh_2(CO)_4Cl_2$ (Yates et al., 1979a).

The other two spectral features (Fig. 3.6), on the other hand, show a shift to higher wave numbers with increasing coverage, typical for intermolecular interaction of CO molecules bonded to extended metal domains (vide supra). The broad band at 1855 cm^{-1} shifts to 1870 cm^{-1}, and the low intensity band at 2056 cm^{-1} shifts to 2070 cm^{-1}, as coverage increases. Based again on the spectra of molecular complexes, these bands were assigned, respectively, to bridging CO, $Rh_2(CO)$, and linear RhCO species on islands or rafts of Rh which are more "crystalline" in character. In fact, Dubois and Somorjai (1980) observed bands at 1870 and 2070 cm^{-1} on the (111) plane of a Rh single crystal at full coverage, but did not see the doublet mentioned above and assigned to isolated $Rh(CO)_2$.

The distribution of the various surface species cannot be obtained from IR data alone, because the extinction coefficients, as well as the integrated absorbances, vary considerably among the different metal$-$CO species (Brown and Darensbourg, 1967). Duncan et al. (1980) have indicated a way to obtain a molar distribution of different surface species by combining the IR results with ^{13}C NMR data. Since the integrated intensity of the NMR spectrum is linearly proportional to the concentration, it is possible to calculate absolute site populations. Although the NMR features originating from different M$-$CO species generally overlap, a line shape analysis permits separation. In the particular case of the system Al_2O_3/Rh/CO, it was found that the ^{13}C NMR signal is composed of contributions of CO groups belonging to two types with two different spin-lattice relaxation times T_1, an order of magnitude apart. T_1 is increased by vicinal nuclei with magnetic moments. Hence, the species with the smaller T_1 was identified with the isolated Rh$-$CO species observed in IRS. The other component was attributed to CO molecules on raft sites and further separated into linear and bridged CO based on chemical shift information. Thus a quantitative distribution of the three species observed in IRS was obtained. In the particular case of a freshly prepared 2.2% Rh on Al_2O_3 sample, exposed to 50 Torr CO at room temperature, the CO site distribution dicarbonyl:linear:bridged was found to be 34:44:22. But this ratio depends strongly on Rh load, Rh precursor, reduction conditions, and support (Worley et al., 1983). Conditions have been found where only the dicarbonyl species is present (0.5% Rh on TiO_2).

As to "mobility" of CO molecules on metal surfaces, several observations might be relevant. Ertl and his school interpreted LEED data, in particular the rapid formation of sharp diffractive spots from CO-covered metal surfaces at room temperature, as indicative of high mobility of the CO molecules on the surface (Conrad et al., 1976). The differential entropy of a CO surface layer, as derived from experimental adsorption isotherms, was also interpreted as showing high surface mobility, and an activation energy of the order of kT (i.e. a few kJ/mol) was estimated for the surface diffusion of CO at low coverage, $\theta \leqq 0.35$ (Behm et al., 1980). Yates and Goodman (1980) compared thermal desorption of CO from a Ni surface with the exchange of surface CO and $C^{18}O$ from the gas phase. Rapid and complete exchange takes place at a temperature (below $0\,°C$), where only a minimal fraction of the CO is separated from the surface in the thermal desorption experiment. Apparently, the CO molecules at the surface easily interconvert among positions of different bonding energy, and the exchange takes place when the CO molecules are in the least bonded situation. Most probably the CO mobility in surfaces is brought about by terminal-bridge bond exchange, as in the molecular carbonyl complexes and clusters.

3.4 CO Dissociation on Metal Surfaces

For almost all transition metals, conditions of temperature and CO pressure can be found, under which the CO molecule dissociates on the metal surface.

The tendency to do so is the greater, the further to the left in a row, and the higher in a column of the Periodic System is the position of the metal. Fig. 3.7 indicates which metal surfaces dissociate CO even at room temperature. The proneness of metal-bonded CO to dissociate has been linked to the increase of the C−O bond length. Although these bond lenghts cannot be measured directly (as in molecular complexes by X-ray crystal structure determination), an indicator has been suggested by Brodén et al. (1976). It is the energy separation $\Delta E (1\pi - 4\sigma)$ between the 1π and 4σ levels as observed in the UPS

IIIB	IVB	VB	VIB	VII	VIII	VIII	VIII	
Sc	Ti	V	Cr	Mn	Fe (3.5)	Co	Ni (3.08)	4
Y	Zr	Nb	Mo	Tc	Ru (3.15)	Rh	Pd (2.9)	5
La	Hf	Ta	W (3.2)	Re	Os	Ir (2,75)	Pt (2.60)	6

Fig. 3.7. Section of the Periodic Table showing the room temperature interaction of CO with the transition metals. Molecular interaction to the right, dissociation to the left of the thick line. The numbers are average values (for different crystal planes) of the energy separation $\Delta E(1\pi - 4\sigma)$ in eV (according to Brodén et al., 1976)

data of CO on metal surfaces (cf. Fig. 3.5; under certain conditions it is possible to separate the 1π from the 5σ level). The molecular orbitals 4σ and 1π of CO are bonding with respect to carbon and oxygen, and not involved in the M−CO bond formation. According to model calculations (Cederbaum et al., 1975), the 1π orbital is considerably more bonding than 4σ. An increase of the carbon-oxygen distance will decrease the orbital overlap, causing both the 4σ and 1π orbitals to be less bonding and hence to have increased energy levels. However, the effect on the strongly bonding 1π is more marked than that on the weakly bonding 4σ. Thus an increase in the energy separation $\Delta E(1\pi - 4\sigma)$ signals an increase in bond length. Values for $\Delta E(1\pi - 4\sigma)$, where known, are included in Fig. 3.7. These numbers clearly increase as one proceeds leftwards in a given row of the Periodic System, or upwards in a given column, in agreement with the experimental evidence for the tendency of the transition metal surfaces to dissociate CO. Brodén et al. (1976) also observed that $\Delta E(1\pi - 4\sigma)$ varies relatively little for different single crystal surfaces or polycrystalline surfaces of the same metal. They concluded that the primary factor determining the occurrence of CO dissociation is the electronic nature of the metal itself, whereas the surface geometry is a second order influence. This is in agreement with the generally accepted bonding picture of CO on transition metal surfaces: electron back donation from a filled metal d orbital into the antibonding CO π^* level should be strongest with the least electronegative metals (towards the left side in each row).

At elevated temperatures, almost all transition metals appear to be able to dissociate carbon monoxide to a lesser or greater extent (exceptions vide infra). When CO dissociation takes place, metal carbide is formed; the oxygen does not escape from the surface, but reacts with a second CO molecule to give CO_2 which is easily desorbed. The whole process, with the overall stoichiometry

$$2CO \rightleftarrows C + CO_2 \qquad\qquad (3.2)$$

is known as carbon monoxide disproportionation, or the Boudouard reaction. Of the two individual steps

$$CO \rightleftarrows C + O \qquad\qquad (3.3)$$

$$CO + O \rightleftarrows CO_2 \qquad\qquad (3.4)$$

the first one is rate determining according to kinetic measurements (Tøttrup, 1976), and the second is rapid. An overall activation energy of ca. 100 kJ/mol has been reported for the Boudouard reaction, in the particular case of CO dissociation on Ni at low CO coverage (Gardner and Bartholomew, 1981). It is generally assumed that the CO molecule is coordinated "end-on" to the surface metal atom prior to dissociation, because this situation, (a) below, is energetically more favorable than that (b) with the CO molecular axis parallel to the surface (Shustorovich et al., 1983):

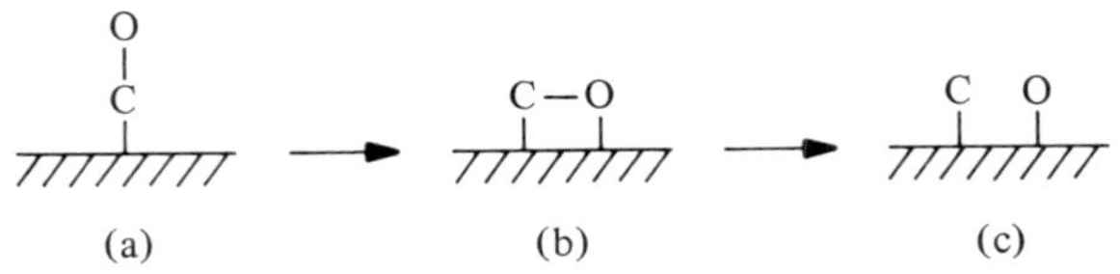

(a)   (b)   (c)

Nevertheless, situation (b) must be visualized as an excited state necessary on the way to complete bond breaking (c).

The amount of CO_2 evolved is frequently used to determine the amount of CO dissociated according to Eq. 3.2, under a given set of conditions (see e.g. Rabo et al., 1978). Other relevant analytical methods include determination of the weight increase by carbide formation, or any of the surface specific spectroscopic methods such as Auger spectroscopy or low energy electron diffraction (LEED).

We shall now review some particular transition metals with regard to their capacity of dissociating carbon monoxide. One of the most studied metals in this respect is nickel. Bahr and Bahr (1928) discovered that *nickel carbide* is formed if fine Ni powder is treated at $250-270\,°C$ with CO gas. They found a weight increase corresponding to the formation of the carbide Ni_3C. At higher temperature, CO in excess of this stoichiometry was dissociated and carbon was deposited onto the nickel carbide. Above $380\,°C$ Ni_3C decomposed to metallic nickel and carbon. If Ni_3C was treated with diluted HCl it dissolved quantitatively without any precipitation of carbon. Bahr and Bahr concluded that the carbon is in fact chemically bonded to the nickel. Excess carbon deposited onto Ni_3C, did not dissolve in HCl.

The crystal structure of Ni_3C was studied by Jacobson and Westgren (1933) using powder X-ray diffraction, and refined by Nagakura (1957) by electron diffraction. The Ni atoms form a close packed hexagonal arrangement and the carbon atoms occupy one third of the octahedral interstices, see Fig. 3.8.

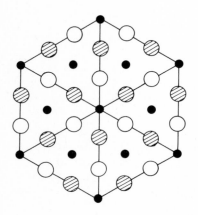

Fig. 3.8. Structure of Ni_3C. The figure shows the projection of the atomic arrangement on a basal plane. Ni atoms (large circles) form a close packed hexagonal array (open and hatched circles represent Ni atoms of two successive planes, respectively). Carbon atoms are located in octahedral interstices (small circles); one third of the interstices is occupied. Lines indicate unit cell; according to Nagakura, 1957

At room temperature, nickel does not dissociate carbon monoxide. This was observed for a single crystal surfaces, under ultrahigh vacuum conditions (Madden and Ertl, 1973), as well as for a silica supported Ni catalyst at ca. 3 atm pressure (Rabo et al., 1978). The coordinatively bonded CO could be removed by thermal desorption, no CO_2 was found. At $300\,°C$, however, carbon monoxide is largely dissociated, When a pulse of CO was sent over a freshly reduced Ni/SiO_2 catalyst, a ratio of C (on the catalyst):CO (on the catalyst):CO_2 (evolved) of ca. $3:1:3$ was found.

Of the two next neighbors of Ni, within the group VIII of the Periodic Table, only *cobalt* behaves very similarly. On *palladium*, on the other hand, CO is molecularly bonded, without any evidence of dissociation even at $300\,°C$ (Rabo et al., 1978). Also on *rhodium* surfaces CO dissociation appears to be practically non existent (Yates et al., 1980, 1982); the upper limit for the probability of dissociation of CO per CO collision with the surface has been estimated to be $\simeq 10^{-8}$, for surface temperatures up to $800\,°C$. It was suggested that earlier findings of enhanced CO decomposition on Rh surfaces containing irregularities such as steps, kinks or other defects (Castner et al., 1981) might have been essentially caused by electron beam damage of CO during Auger spectroscopic measurements. This view is corroborated by a field emission spectroscopic investigation of the interaction of CO and Rh. A field emission tip exposes both smooth and open (stepped) planes of a given metal crystal. No CO dissociation has been detected on a Rh tip at CO pressures up to 10^{-1} Torr, and at surface temperatures up to 700 °C (Gorodetskii and Nieuwenhuys, 1981).

In the case of *ruthenium*, some dissociation is observed at elevated temperatures; CO_2 starts to form in minor amounts at $\simeq$ 140 °C, with a maximum at $\simeq$ 470 °C, (Low and Bell, 1979). To explain isotopic exchange between $^{12}C^{18}O$ and $^{13}C^{16}O$ at 100 °C, in the complete absence of CO_2 and hence, presumably, of CO dissociation, Bossi et al. (1980) suggested that CO in a dicarbonyl species, on a partially oxidized ruthenium center, might be bonded loosely enough (less back donation), and also geometrically suited, to react with a contiguous similar CO group in a four-center exchange reaction. McCarty and Wise (1979), on the other hand, favor a reversible CO dissociation involving CO bond rupture for the same isotopic exchange reaction, observed at $T \cong$ 100 °C, although the CO_2 formation is minor. This particular question appears unresolved at present.

Carbon monoxide dissociation on *iron* occurs slowly at room temperature, but proceeds rapidly at 450 °C (Kishi and Roberts, 1975). Depending on time and temperature of interaction, different iron carbides are formed: θ-Fe_3C, χ-Fe_2C_2 (Hägg carbide), ε-Fe_2C, ε'-$Fe_{2.2}C$ (see e.g. Raupp and Delgass, 1979).

In those cases where carbides are formed by the interaction of CO with a group VIII metal, these carbides react easily with hydrogen to give mainly methane, and small amounts of higher hydrocarbons. This reaction, which was already observed by Bahr and Bahr (1928), will be discussed in Chap. 4.1 (non catalytic) and 7.1 (catalytic). Carbides of the transition metals further to the left side in the Periodic Table have very stable metal−C bonds. Thus, whereas the heat of formation of iron, cobalt and nickel carbide is + 21, + 40, and + 46 kJ/g-atom C, respectively, that of TiC, ZrC, and Mo_2C is − 226, − 188 and − 48 kJ/g-atom C, respectively (Ponec and van Barneveld, 1979). These latter three carbides do not react with hydrogen even at 300 °C (Kojima et al., 1980).

Finally, it should be mentioned that the metal carbides on surfaces have also their "models" in molecular transition metal cluster compounds. It has been demonstrated, for instance, that $Ru_3(^{13}CO)_{12}$ undergoes a thermal reaction in the presence of $Mn(CO)_5^-$, whereby a CO ligand is transformed into a carbidic ligand (Bradley et al., 1980):

$$Ru_3(^{13}CO)_{12} \xrightarrow{[Mn(CO)_5]^-} [Ru_6{}^{13}C(^{13}CO)_{16}]^{2-}.$$

Other polynuclear carbide cluster compounds are known, but generally the source of the carbide is not CO (see e.g. Albano et al., 1973). In some clusters the carbide carbon is located in the center of the metal core, as for instance, in the octahedral core of $[Fe_6C(CO)_{16}]^{2-}$ (Churchill et al., 1971); in others it is peripheral with regard to the metal core as in $[Fe(\mu_4$-$C)(CO)_{12}]^{2-}$ (see Fig. 3.9; the distorted tetrahedral Fe_4 structure has been named a "butterfly"

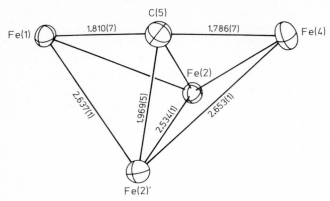

Fig. 3.9. Fe_4C core cluster structure for $[Fe_4(\mu_4\text{-}C)(CO)_{12}]^{2-}$, "butterfly cluster"; each iron atom has three terminal carbonyl ligands (Davis et al., 1981)

cluster). In general, the carbide carbon in these clusters appears to have a low degree of activity towards hydrogen (Tachikawa et al., 1980). But the butterfly complex represented in Fig. 3.9 reacts with Ag^+ and H_2, under mild conditions to give the protonated complex $HFe_4(\mu^2\text{-}CH)(CO)_{12}$ (Davis et al., 1981):

$$[Zn(NH_3)_4]^{2+}[Fe_4C(CO)_{12}]^{2-} \xrightarrow[H_2]{Ag^+} HFe_4(\mu^2\text{-}CH)(CO)_{12}. \qquad (3.5)$$

This interesting reaction is assumed to proceed via oxidation of the starting complex to the transitory and coordinatively unsaturated $Fe_4C(CO)_{12}$. If the oxidation of the starting complex was carried out in the presence of CO instead of H_2, the neutral $Fe_4C(CO)_{13}$ could be isolated. The crystal structure of this latter was determined by Bradley et al. (1981) (who prepared it in a different way, starting from the octahedral $[Fe_6C(CO)_{16}]^{2-}$ cluster). It is similar to that represented in Fig. 3.9, except that the extra CO ligand bridges Fe(2) and Fe(2)'.

3.5 Conclusion

Carbon monoxide interacts with transition metals either as a molecule, or it dissociates giving a metal carbide and oxygen which desorbs as CO_2. In the overwhelming majority of the known carbonyl complexes and clusters, only molecular bonding is present, whereby the carbon atom of the CO molecule is bonded to one, two, or three metal atoms, in a linear, doubly or triply bridging fashion, respectively. *The molecular bonding of CO on metal surfaces is similar to that in complexes and clusters,* as indicated by bond energies, bond length, photoelectron spectra and CO stretching frequencies. But dissociative adsorption of CO on metal surfaces does also take place, in particular at elevated temperatures and the question arises whether one or the other of the two forms (molecular of dissociated), or both, are involved in the catalytic hydrogenation reactions under consideration in this book. This question will be amply discussed in the following Chapters.

3.6 References

Albano VG, Sausoni M, Chini P, Martinengo S (1973) J Chem Soc Dalton, 651
Alich A, Nelson NH, Strope D, Shriver DF (1972) Inorg Chem 11:2976
Andersson S (1977) Solid State Commun 21:75
Baerends EJ, Ros P (1975) Mol Phys 30:1735
Bahr HA, Bahr TH (1928) Ber Dtsch Chem Ges 61:2177
Basolo F, Pearson RG (1967) Mechanism of Inorganic Reactions. 2. Ed. J. Wiley, New York
Beach NA, Gray HB (1968) J Amer Chem Soc 90:5713
Behm RJ, Christmann K, Ertl G, van Hove MA (1980) J Chem Phys 73:2984
Behrens RG (1978) J Less Common Met 58:47
Benfield RE, Johnson BFG (1981) Transition Met Chem 6:131
Blyholder G (1964) J Phys Chem 68:2772
Bossi A, Carnisio G, Garbassi F, Giunchi G, Petrini G (1980) J Catal 65:16
Bradley JS, Ansell GB, Hill EW (1980) J Organometal Chem 184:C33
Bradley JS, Ansell GB, Leonowicz ME, Hill EW (1981) J Amer Chem Soc 103:4968
Braterman PS (1972) Struct Bonding 10:57
Braterman PS (1976) Struct Bonding 26:1
Brodén G, Rhodin TN, Brucker C (1976) Surface Sci 59:593
Brown TL, Darensbourg DJ (1967) Inorg Chem 6:971
Bullet DW, O'Reilly EP (1979) Surface Sci 89:274
Calderazzo F, Fachinetti G, Marchetti F (1981) J Chem Soc Chem Commun 181
Castner DG, Dubois LH, Sexton BA, Somorjai GA (1981) Surface Sci 103:L134
Cederbaum LS, Domcke W, von Niessen W, Brenig W (1975) Z Physik 21:381
Chini P (1970) Pure Appl Chem 23:489
Chini P, Martinengo S, McCaffrey DJA, Heaton BT (1974) J Chem Soc Chem Commun 310
Chini P, Longoni G, Albano VG (1976) Adv Organometal Chem 14:285
Churchill MR, Wormold J, Knight J, Mays MJ (1971) J Amer Chem Soc 93:3073
Colton R, McCormick MJ (1980) Coord Chem Rev 31:1
Commons CJ, Hoskins BF (1975) Austr J Chem 28:1663
Connor JA (1977) Topics Curr Chem 71:71
Conrad H, Ertl G, Knözinger H, Küppers J, Latta EL (1976) Chem Phys Lett 42:115
Conrad H, Ertl G, Küppers J, Latta EL (1977) Proc 6th Internat Congr Catalysis (Discussion)
Cotton FA (1976) Progress in Inorg Chem 21:1
Cotton FA, Wilkinson G (1972) Advanced Inorganic Chemistry. Interscience, New York, 3rd Ed.
Davis JH, Beno MA, Williams JM, Zimmie J, Tachikawa M, Muetterties EL (1981) Proc Natl Acad Sci USA 78:668
Dubois LH, Somorjai GA (1980) Surface Sci 91:514
Duncan TM, Yates JT, Vaughan RW (1980) J Chem Phys 73:975
Egger KW, Cocks AT (1973) Helv Chim Acta 56:1516
Engel T, Ertl G (1979) Adv Catal 28:1
Gardner DC, Bartholomew CH (1981) Ind Eng Chem Prod Res Dev 20:80
Gorodetskii VV, Nieuwenhuys BE (1981) Surface Sci 105:299

Heinz K, Lang E, Müller K (1979) Surface Sci 87:595
Henrici-Olivé G, Olivé S (1977) Coordination and Catalysis. Verlag Chemie, Weinheim, New York
Hillier IH, Saunders VR (1971) Mol Phys 22:1025
Jacobson B, Westgren A (1933) Z phys Chem 20:361
Johnson BFG (ed) (1980) Transition Metal Clusters. Wiley, New York
Kishi K, Roberts MW (1975) J Chem Soc Faraday I 71:1715
Kojima I, Miyazaki E, Yasumori I (1980) J Chem Soc Chem Commun 573
Koridze AA, Kizas OA, Astakhova NM, Petrovich PV, Grishin YuK (1981) J Chem Soc Chem Commun 853
Kroeker RM, Hausma PK (1981) Catal Rev-Sci Eng 23:553
Kroeker RM, Hausma PK, Kaska WC (1980) J Chem Phys 72:4845
Low GG, Bell AT (1979) J Catal 57:397
McCarty JG, Wise H (1979) Chem Phys Lett 61:323
Madden HH, Ertl G (1973) Surface Sci 35:211
Manassero M, Sansoni M, Longoni G (1976) J Chem Soc Chem Commun 919
Mond LC, Langer C, Quinke F (1890) J Chem Soc 57:749
Nagakura S (1957) J Phys Soc, Japan 12:482
Nieuwenhuys BE (1981) Surface Sci 105:505
Orchin M, Jaffé HH (1971) Symmetry Orbitals and Spectra. Wiley-Interscience, New York, Chapt. 6
Pettifor DG (1977) J Phys F 7:613
Politzer P, Kammeyer CW, Bauer J, Hedges WL (1981) J Phys Chem 85:4057
Ponec V, van Barneveld WA (1979) Ind Eng Chem Prod Res Dev 18:268
Rabo JA, Risch AP, Poutsma ML (1978) J Catal 53:295
Raupp GB, Delgass WN (1979) J Catal 58:348
Shustorovich E, Baetzold RC, Muetterties EL (1983) J Phys Chem 87:1100
Smith DW (1976) J Chem Soc Dalton 834
Somorjai GA (1981) Chemistry in Two Dimensions: Surfaces. Cornell University Press, Ithaca, NY
Tachikawa M, Sievert AC, Muetterties EL, Thompson MR, Day CS, Day VW (1980) J Amer Chem Soc 102:1725
Tottup PB (1976) J Catal 42:29
Wender I, Pino P (eds) (1977) Organic Synthesis via Metal Carbonyls. Wiley, Vol. 2
White JM (1983) J Phys Chem 87:915
Woolley RG (1980) Platinum Met Rev 24:26
Worley SD, Mattson GA, Caudill R (1983) J Phys Chem 87:1671
Yang AD, Garland CW (1957) J Phys Chem 61:1504
Yates JT, Goodman DW (1980) J Chem Phys 73:5371
Yates JT, Kolasinski K (1983) J Chem Phys 79:1026
Yates JT, Duncan TM, Worley SD, Vaughan RW (1979a) J Chem Phys 70:1219
Yates JT, Duncan TM, Vaughan RW (1979b) J Chem Phys 71:3908
Yates JT, Williams ED, Weinberg WH (1980) Surface Sci 91:562
Yates JT, Williams ED, Weinberg WH (1982) Surface Sci 115:L93

4 Non-Catalytic Interaction of CO with H_2

4.1 On Metal Surfaces

A first insight into mechanistic aspects of the hydrogenation of carbon monoxide may be obtained by studying the coadsorption of CO and H_2 under non-catalytic conditions, in particular at low pressure. Only the very first stages of the hydrogenation process(es) are amenable to observation under these conditions, since thermodynamics favor only one-carbon species at low pressure.

A major difficulty in such endeavor is the limited detectability of surface species. No easy direct methods are available, but a number of indirect ways have been designed to estimate the behavior of coadsorbed CO and H_2. Volumetric coadsorption measurements have shown that, at temperatures between 50 and 100 °C and for different metal surfaces such as Fe, Co, Ni, and Ru, the total adsorption from a CO/H_2 mixture is greater than the sum of the components each adsorbed separately (for a review see Vannice, 1976). It was concluded that some kind of a surface complex is formed. In many instances the coadsorbed ratio CO/H_2 was near unity and it was deduced that the surface compound might be adsorbed formaldehyde, but in other cases it was shown that the CO/H_2 ratio in the adsorbed phase was dependent on temperature and pressure. Under ultrahigh vacuum conditions the interaction of H_2 and CO is sometimes less clear; thus in the case of Ru enhancement was only found if CO was adsorbed onto preadsorbed H_2 (Kraemer and Menzel, 1975), and at $- 173$ °C it was even found that preadsorbed CO strongly blocks H_2 adsorption, and preadsorbed H_2 blocks CO adsorption, though less strongly (Peebles et al., 1982). Ultrahigh vacuum conditions ($< 10^{-10}$ Torr) are, evidently, opposed to the formation of surface species containing both, CO and H_2. Although an ordered coadsorbate structure was found under these conditions (LEED data, Conrad et al., 1977), no evidence whatsoever was found for a surface complex.

Some more information comes from thermal desorption spectroscopy (TDS). Craig (1982) investigated the coadsorption of H_2 and CO on a recrystallized platinum ribbon. Figure 4.1 shows TDS data for increasing exposure (in L = Langmuir = 10^{-6} Torr $\cdot$ s) of the platinum sample to a H_2/CO gas mixture at a ratio $11:1$; H_2 (mass 2) and CO (mass 28) are monitored as a function of desorption temperature in Figs. 4a and 4b respectively. The H_2 desorption spectra show a low temperature peak at ca. 160 °C (433 K) which is

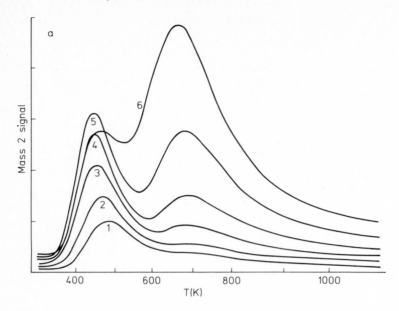

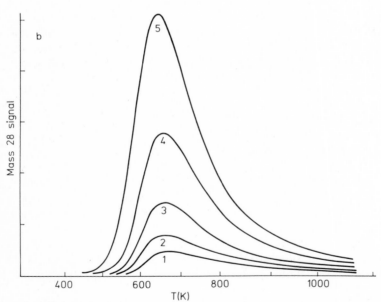

Fig. 4.1 a, b. Mass 2 (**a**) and mass 28 (**b**) thermal desorption spectra for coadsorption of H_2 and CO on Pt, at room temperature; gas mixture ratio $H_2/CO = 11$; (1)−(6) exposures (in L) of 0.05, 0.10, 0.20, 0.40, 0.80, 1.6. Exposure pressure 5×10^{-9} Torr (Craig, 1982). Reproduced by permission of North-Holland Physics Publishing

characteristic for the desorption of pure H_2 from the Pt surface under the same conditions. There is, however, an additional peak at ca. 400 °C (680 K), which increases with exposure. The corresponding CO spectra show only one regularly increasing peak. A comparison of Figs. 4a and 4b indicates that the high temperature mass 2 peak increases proportionally to the quantity of CO present on the surface. No signals corresponding to mass numbers other than 2 and 28 were observed. Since the appearance and the dramatic increase of the high temperature mass 2 peak with exposure is only observed in the presence of CO, it is assumed that it is caused by a surface complex formed by the interaction of CO and H_2, possibly HCO (formyl). This would imply the supposition that this surface complex decomposes at the temperature of CO desorption which is about the same as that of the high temperature mass 2 peak. Similar evidence for the formation of some kind of a surface complex between adsorbed CO and H_2 was found for W, Ni, Ru, and Fe surfaces (Benziger and Madix, 1982, and references therein).

Surface infrared spectroscopy has also been used to study coadsorption. Primet and Sheppard (1976) found that ν_{CO} of adsorbed CO (on Ni, 25 °C) is increased from 2040 cm^{-1} (CO alone) to 2070 cm^{-1} in the presence of H_2. This was interpreted as the result of an electron-attracting action of the dissociated hydrogen on the surface, thus decreasing the electron back-donation into the π^* CO orbital. In terms of surface complexes this appears to imply the formation of a Ni(H)CO species. On a Ni(111) surface, an "oxyhydrocarbonated" species has been detected after coadsorption of H_2 and CO at 140 °C; it was characterized by vibrations at 2920, 1436 and 1355 cm^{-1}, and by simultaneous desorption of hydrogen and carbon monoxide at 180 °C (Bertolini and Imelik, 1979).

Similar observations on a Rh surface were reported by Worley et al. (1983). These authors had found that on a 0.5% Rh/TiO$_2$ surface only one species, namely the geminal dicarbonyl was present after CO exposure (cf. Table 3.6; on the TiO$_2$ carrier the doublet appears at 2095 and 2026 cm^{-1}). If the surface was exposed to CO/H$_2$ at room temperature, the spectrum did not change, but after heating for 4h at 210 °C, the doublet disappeared, and a new band formed at 2043 cm^{-1}. With regard to the ν_{sym} (= 2095 cm^{-1}) of the dicarbonyl, this is a clear decrease, indicating a weakening of the C—O bond, as it may be expected for a replacement of one of the CO ligands by a hydride ligand (less acceptor than CO, hence more π-back donation into π^* of the remaining CO ligand). Hence, the authors attributed the spectral change to the transition

$$
\begin{array}{ccc}
\text{OC} \diagdown \quad \diagup \text{CO} & & \text{H} \diagdown \quad \diagup \text{CO} \\
\quad \text{Rh} & \longrightarrow & \quad \text{Rh}
\end{array}
$$

Saussey et al. (1983) investigated the coadsorption of CO and H_2 on Cu—ZnO (catalyst for the synthesis of methanol from CO and H_2) by IR spectroscopy. While the ν_{CO} region was cluttered by carbonate species, they were able to detect a doublet at 2770 and 2661 cm^{-1} which was replaced by a single band at 2020 cm^{-1} in the case of D_2/CO adsorption. It was suggested that the doublet is due to the $\nu(CH)$ vibration of a formyl group, which generally gives rise to two bands caused by a strong Fermi resonance between the $\nu(CH)$ and $2\delta(CH)$ levels. This perturbation does not affect the $\nu(CD)$ mode which explains the appearance of a single band in the case of the $-CDO$ group.

Deluzarche et al. (1978) have provided chemical evidence for the presence of formyl groups on an unsupported Ni surface, after treatment with CO and H₂ at normal pressure. Instead of desorbing the surface compound, these authors trapped it with $(CH_3)_2SO_4$, $(CH_3CH_2)_2SO_4$ or CH_3I. Gas-chromatographic detection of CH_3CHO, CH_3CH_2CHO, and CH_3CHO, respectively, indicated the presence of formyl groups, $-CHO$, on the surface, prior to trapping.

It follows then that hydridocarbonyl species, and in some cases formyl complexes, are formed on transition metals if CO and H₂ are coadsorbed, under suitable conditions. These species have their counterparts in the field of molecular transition metal complexes (cf. Sect. 4.2).

Carbon deposited on a metal surface by the Boudouard reaction (cf. Eq. 3.2) reacts with hydrogen even at room temperature, to give methane (see, e.g. Rabo et al., 1978). Using temperature programming techniques, it was shown in the cases of Ni (McCarty and Wise, 1979) and Fe (Kieffer and van der Baan, 1982) that more than one type of carbon species can be deposited on the surface by decomposition of CO. After the deposition the catalyst was purged with inert gas, and hydrogen was added. The catalyst bed temperature was then slowly increased from room temperature to $> 400\,°C$ while monitoring the effluent gas composition, either by mass spectroscopy or by gas chromatography (temperature-programmed surface reaction, TPSR). It was found that methane is formed in different stages as the temperature is raised. In the case of Ni, two major carbon species were observed based on their reactivity towards H₂ (see Fig. 4.2): a more reactive form (α-carbon) assumed to consist of individual isolated $Ni-C$ centers, with a peak temperature of $\simeq 200\,°C$ (470 K), and a less reactive form (β-carbon) taken as amorphous carbon, with a maximum at $\simeq 400\,°C$. The ratio of the two forms remained approximately constant with increasing carbon deposit.

Under certain conditions, bulk carbide is formed (see Sect. 3.4). Bahr and Bahr (1928) have shown that bulk nickel carbide (Ni_3C) reacts with hydrogen, at relatively low temperatures to give methane and some unidentified higher hydrocarbons. Galwey (1962) restudied this reaction and found methane and ethane in a molar ratio 20:1, constant over a large range of Ni_3C concentration, and independent of the temperature in the range $250-300\,°C$. The kinetics were described by

$$\frac{d[CH_4]}{dt} = k\, p_{H_2}[C] \tag{4.1}$$

where p_{H_2} is the hydrogen pressure and [C] is the concentration of carbon in the sample. An activation energy of $\simeq 17\,kJ/mol$ was determined for the formation of CH_4, and $\simeq 29\,kJ/mol$ for C_2H_6. It was suggested that carbon

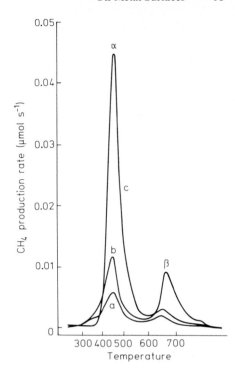

Fig. 4.2. Temperature programmed surface reaction with H_2 following carbon deposition by exposure of Ni/Al$_2$O$_3$ to CO at $\simeq 280\,°C$; a, b, c: increasing carbon deposit; α, β: α- and β-carbon (McCarty and Wise, 1979). Reproduced by permission of Academic Press

atoms in the bulk are in equilibrium with those on the surface. The surfacial carbon reacts with H_2, and is replaced by C diffusion from the interior. McCarty and Wise (1979) reported that bulk carbide is slightly less reactive towards hydrogen than the isolated surface Ni$-$C species (under the same conditions as those of Fig. 4.2, bulk carbide has a peak temperature for CH_4 formation of ca. 280 °C). This may be due to the preequilibrium mentioned above.

At elevated temperatures ($> 400\,°C$) nickel carbide decomposes into its atomic components, and amorphous carbon deposits transform to crystalline graphitic carbon, which does not react with hydrogen. This temperature dependence has been shown very nicely by Goodman et al. (1980), with the aid of Auger electron spectroscopy. A Ni crystal (100) was exposed to 24 Torr CO at 327 °C or at 427 °C. The AES spectra shown in Fig. 4.3 a and b were observed, respectively. Comparison with the AES carbon signals from Ni$_3$C and single crystal graphite (Figs. 4.3 c and d) indicated that at the lower temperature carbidic carbon was deposited (it reacted with H_2 to give CH_4), whereas at the higher temperature graphitic carbon was formed, this latter species did not disappear in H_2, even after prolonged heating at 380 °C.

Under non graphitizing conditions, large amounts of amorphous carbon can be deposited, in particular on Fe, Co and Ni. The metal is fragmented by the carbon, and metal particles are found in the accumulated carbon (Rostrup-Nielsen and Trimm, 1977). In the particular case of Ni, it was shown that such

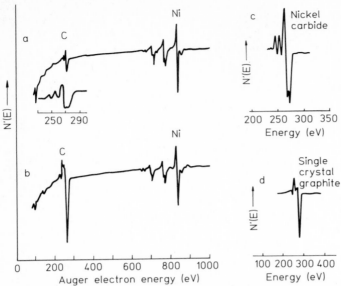

Fig. 4.3a – d. AES spectra of Ni(100) crystallographic plane, following exposure to 24 Torr CO at 327 °C (**a**) and at 427 °C (**b**), and comparison with AES carbon signals of nickel carbide (**c**) and single crystal graphite (**d**) (Goodman et al., 1980). Reproduced by permission of Academic Press

carbon reacts between 300 and 450 °C with H_2 to give methane; if, however, the nickel particles were washed with hydrochloric acid, no methane formation was observed up to 600 °C, indicating the necessity of the presence of a metal-carbon interphase for the reaction to take place (Audier et al., 1979). Obviously, Ni is necessary to activate (dissociate) the hydrogen, but also the Ni–C bond needs to be present. In a very limited way, the reaction of carbidic carbon on metal surfaces has also a counterpart in molecular complexes (see Sect. 4.2.5).

4.2 CO/H_2 Interaction in Transition Metal Complexes

4.2.1 Hydridocarbonyl Complexes

Transition metal complexes containing simultaneously carbonyl and hydride ligands are known in large number (see e.g. Muetterties, 1972); several of them have been already mentioned in Tables 2.1 and 2.2. In many instances the carbonyl-hydride complexes form if carbonyl cluster compounds are reacted with hydrogen. In the case of cobalt the reaction:

$$Co_2(CO)_8 \xrightarrow{H_2} 2\,HCo(CO)_4 \qquad (4.2)$$

takes place under relatively mild conditions (100 atm and 50 °C), while more drastic conditions are required for the formation of $HIr(CO)_4$ from $Ir_4(CO)_{12}$ (300−400 atm, 20−200 °C) and even more so for the formation of $HRh(CO)_4$ from $Rh_4(CO)_{12}$ (> 1500 atm, $CO:H_2 = 4.5:1$) (Vidal and Walker, 1981, and references therein).

Some cases are known where hydrogen reversibly displaces carbon monoxide under mild conditions (ambient temperature and pressure), e.g.

$$Ir(CO)_3(PPh_3)_2 + H_2 \rightleftarrows Ir(H_2)(CO)_2(PPh_3)_2 + CO \qquad (4.3)$$

(Church et al., 1970), or

$$[Pt_2H_2(\mu\text{-}H)(\mu\text{-}dpm)_2]^+ + CO \rightleftarrows [Pt_2H(CO)(\mu\text{-}dpm)_2]^+ + H_2 \qquad (4.4)$$

(dpm = $Ph_2PCH_2PPh_2$; Fischer et al., 1982).

4.2.2 Formyl Complexes

Formyl complexes, which are assumed to be among the surface species formed on transition metal surfaces under the influence of CO and H_2 (Sect. 4.1), are today a firm reality in molecular coordination chemistry. The first such complex was prepared by Collman and Winter (1973), who obtained the anionic complex [(CO)₄FeCHO]⁻ from $Na_2Fe(CO)_4$ and acetic formic anhydride:

$$Na_2Fe(CO)_4 + \overset{O}{\overset{\|}{HC}}-O-\overset{O}{\overset{\|}{C}}CH_3 \rightarrow Na[H\overset{O}{\overset{\|}{C}}Fe(CO)_4] \; . \qquad (4.5)$$

The formyl group was characterized by NMR ($\delta = 14.95$ for the formyl H) and IR ($\nu_{CO} = 1577 \; cm^{-1}$). The formyl complex was found to be relatively stable in solution; at 25 °C it decomposed slowly ($t_{1/2} \gtrsim 12$ days) to [HFe(CO)₄]⁻. It was not possible to prepare the complex directly by the forward reaction of equilibrium (4.6):

$$[HFe(CO)_4]^- + CO \rightleftarrows [H\overset{O}{\overset{\|}{C}}Fe(CO)_4]^- \qquad (4.6)$$

and it was assumed that this equilibrium lies far to the left.

Other routes to formyl complexes have been found later, in particular by Casey and Gladysz and their coworkers (see e.g. Casey et al., 1980; Gladysz, 1982). In most cases metal carbonyls were treated with trialkyl- or trialkoxy-borohydrides, according to the simplified equation

$$HBR_3 + L_xMCO \rightarrow BR_3 + L_xM\overset{\displaystyle O}{\overset{\|}{C}}H \tag{4.7}$$

with L=PPh₃, P(OPh)₃, CO. Most formyl complexes prepared by this route are anionic, but neutral complexes have also been synthesized, e.g. by reducing carbonyl cations with LiHBR₃:

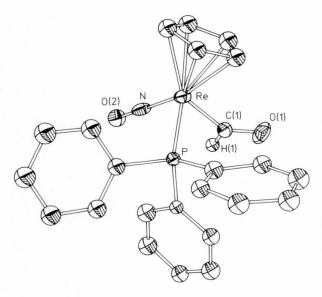

$$\tag{4.8}$$

(Cp = η⁵-C₅H₅, L = PPh₃). Most of the reported formyl complexes were unstable and could not be isolated, though their IR and NMR spectra were recorded. In a few cases, however, the complexes were stable enough for isolation and single crystal X-ray structural analysis. Thus, the molecular structure of the rhenium-formyl complex of Eq. (4.8) is represented in Fig. 4.4. An amazingly short rhenium-formyl carbon bond distance of 205.5 pm was measured, and interpreted in terms of considerable electron back donation (partial Re−C double bond; destabilization of the formyl C−O bond); the low

Fig. 4.4. Molecular structure of [η⁵-C₅H₅)RePPh₃(NO)(CHO)] (Wong et al., 1979). Reproduced by permission of The Royal Society of Chemistry

stretching frequency $v_{CO} = 1588$ cm^{-1} (as compared to $v_{CO} = 1744$ for, e.g., formaldehyde) is in agreement with this interpretation. The same type of bonding of the formyl ligand was found by Casey et al. (1980) for the complex [(RO)$_3$P(CO)$_3$Fe(CHO)], for which X-ray structural analysis revealed an even shorter metal-formyl carbon bond of 195 pm.

A different route to formyl complexes was reported by Thorn (1982): by oxidative addition of the C−H bond of formaldehyde to a square-planar Ir(I) complex. The reaction (Eq. 4.9) is assumed to involve the intermediate formation of a complex containing a π-bonded formaldehyde molecule (such species are well documented in other cases, see Sect. 4.2.3).

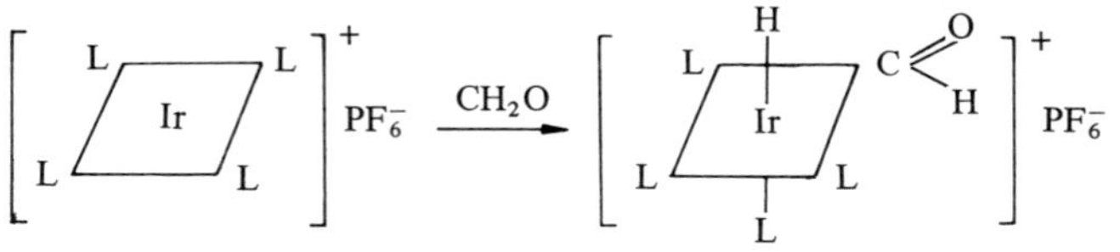

$$(4.9)$$

(L = P(CH$_3$)$_3$). Hydrido and formyl ligands are in cis positions; an IR band at 1600 cm^{-1} has been attributed to v_{CO}. The neutral complexes

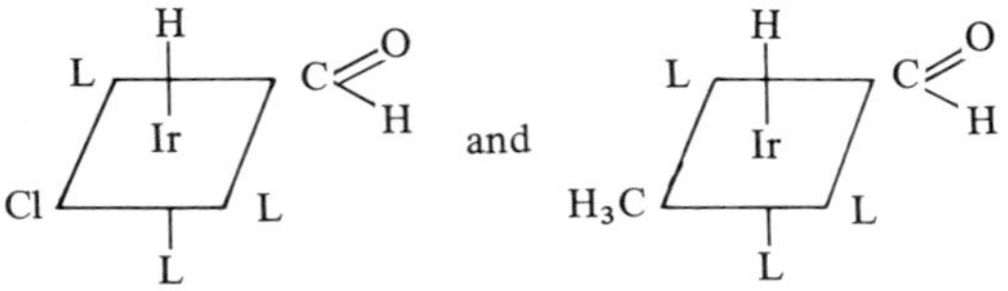

have been prepared by the same route.

The direct interaction of a hydride ligand with CO had eluded observation until very recently, although Basolo and Pearson (1967) suggested intermediate hydride migration and formyl formation in order to explain an anomalously rapid exchange of HMn(CO)$_5$ with ^{14}CO. Byers and Brown (1977) supported this assumption, using ^{13}CO and analyzing (IR) the various isomeric species resulting from this reaction (Eq. 4.10). Although the postulated formyl complex was not observed directly, the distribution of those isomers which retained label after decarbonylation made it highly probable that it had been formed as an intermediate: the cis isomer should be twice as abundant as the trans; in fact a ratio 1:0.48 was found.

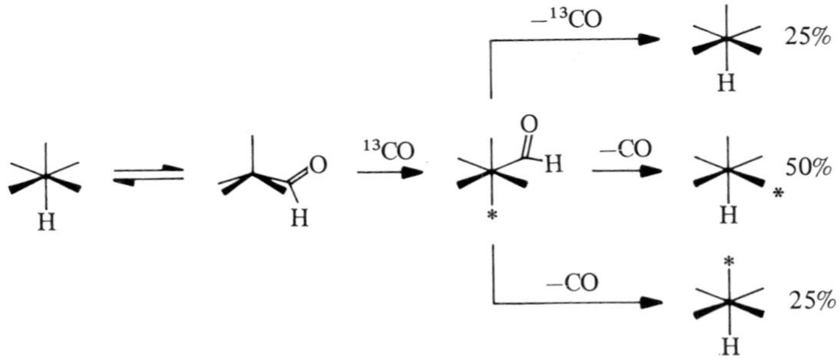

$$(4.10)$$

Formyl complex formation by intermolecular hydride transfer from an anionic ruthenium hydrido complex to a cationic Re carbonyl species was reported by Dombek and Harrison (1983). The reaction proceeds rapidly at $25\,^{\circ}C$ according to the suggested stoichiometry:

$$3[HRu(CO)_4]^- + 2[CpRe(CO)_2(NO)]^+$$
$$\rightarrow [HRu_3(CO)_{11}]^- + 2[CpRe(CO)(NO)(CHO) + CO.$$

No detailed mechanism has been suggested.

Thermodynamically stable formyl complexes of rhodium porphyrins have been reported by Wayland et al. (1983). Rhodium octaethylporphyrin dimer $[Rh(OEP)]_2$ reacts with hydrogen and carbon monoxide in a two-step process, according to:

$$[Rh(OEP)]_2 + H_2 \quad \rightleftarrows 2\,Rh(OEP)(H) \tag{4.11}$$

$$Rh(OEP)(H) + CO \rightleftarrows Rh(OEP)(CHO) \tag{4.12}$$

The formyl complex was characterized by ^1H and ^{13}C NMR, showed an IR absorption band at $1700\ cm^{-1}$ due to the formyl CO, and could be crystallized. A different synthesis (applicable also to other rhodium porphyrin complexes) starts from Rh(porphyrin)(Cl)(CO) which is reacted with KOH and CO. The hydride is formed as an intermediate, so that the formyl complex again arises from Eq. (4.12). In view of the rigid structure of the quadridentate porphyrin ligand, Wayland et al. assumed that formyl formation takes place without prior coordination of the CO molecule. Since this assumption does not appear to fit into the known framework of CO coordination chemistry, another pathway may be suggested. Ogoshi et al. (1975) described the alkylation of Rh(I) by transfer of an alkyl group originally bonded to one of the nitrogen atoms of a porphyrin ligand to an axial position of the complex, accompanied by the closure of the equatorial N—metal bond. Along similar lines the hydrogen ligand (cf. Eq. 4.11) may migrate from one axial position to the other via a N of the equatorial plane, and generate the formyl ligand by insertion of a coordinated CO ligand.

A vanadium formyl complex was obtained according to

$$Cp_2'V-H + 2\,CO \rightarrow Cp_2'V {\displaystyle \mathop{\textstyle <}_{CO}^{CHO}} \tag{4.13}$$

$[Cp' = \eta^5\text{-}C_5(CH_3)_5]$ (Floriani, 1983). The formyl complex was isolated in form of stable crystals; formyl and carbonyl ligand exhibit IR bands at $v_{CO} = 1690$ and $1865\ cm^{-1}$, respectively.

Cameron et al. (1983) observed that the oxidation of a hydridoiron complex with Cu(II), in alcohol, leads to the corresponding formate and suggested the formation of an intermediate formyl complex as the most plausible explanation:

$$CpFe(CO)_2H + Cu(II) \rightarrow [CpFe(CO)_2H]^+ + Cu(I)$$

$$[CpFe(CO)_2H]^+ + ROH \rightarrow [CpFe(CO)(CHO)(ROH)]^+ \rightarrow HCOOR$$

Decrease of the metal-to-ligand back donation or a lowering of the energy of the metal$-$CO σ bond orbitals were discussed as possible reasons for the facilitated hydrogen migration in the Fe(III) complex.

The formation of a dihapto (or π-bonded) formyl ligand was reported by Fagan et al. (1981). A hydridothorium complex was exposed to CO, in toluene solution at $-78\,^\circ$C:

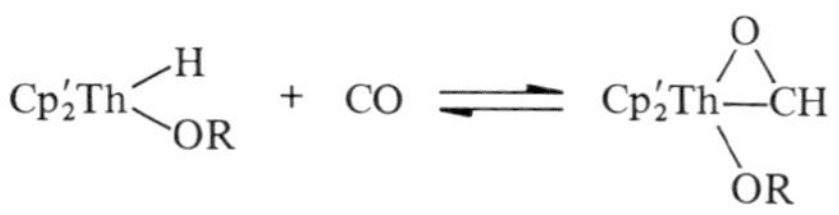

$$Cp'_2Th{\Large\langle}^H_{OR} + CO \rightleftharpoons Cp'_2Th\!\!-\!\!\overset{\displaystyle O}{\overset{/\backslash}{C}}H \qquad (4.14)$$
$$\qquad\qquad\qquad\qquad\qquad |$$
$$\qquad\qquad\qquad\qquad\qquad OR$$

(Cp$'$ = η^5-C$_5$(CH$_3$)$_5$; R = C[C(CH$_3$)$_3$]$_2$H). Elimination of excess CO reversed the reaction. The π formyl complex was characterized by ^{1}H and ^{13}C NMR (comparison with known π-acyl thorium complexes). Presumably the acidic nature of the Th(IV) species favors the π-coordination of the formyl ligand (as a two-electron donor) over the σ-coordination (one-electron donor) observed in most other formyl complexes (cf. Fig. 4.4).

A bridging formyl ligand, with dihapto coordination to two metal centers was found by Belmonte et al. (1983) in the case of a tantalum complex:

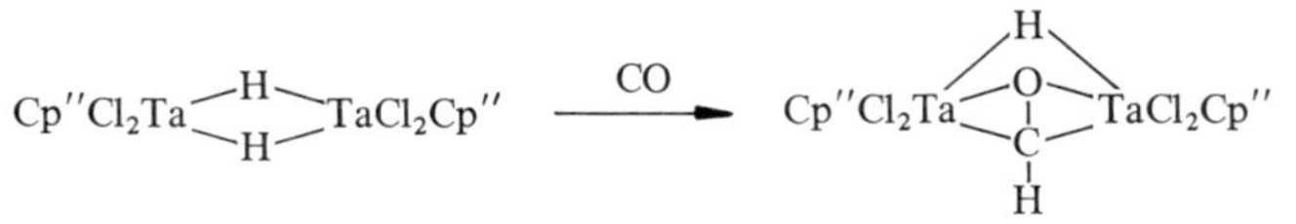

$$Cp''Cl_2Ta{\Large\langle}^H_H TaCl_2Cp'' \xrightarrow{\ CO\ } Cp''Cl_2Ta \underset{\underset{H}{\overset{|}{C}}}{\overset{\overset{H}{\diagup}\diagdown}{\underset{\diagdown}{O}}} TaCl_2Cp''$$

[Cp$''$ = η^5-C$_5$(CH$_3$)$_4$(C$_2$H$_5$)]. The symmetric bridging position of the formyl ligand has been determined by X-ray structural analysis of the crystalline complex. Again, the acidic character of the quadrivalent transition metal species may favor this type of bonding.

4.2.3 Formaldehyde Complexes

Formaldehyde has been mentioned in Sect. 4.1 as one of the possible surface species formed from CO and H$_2$ on transition metal surfaces. Again, this

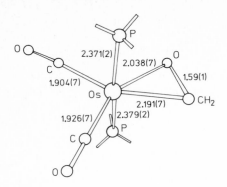

Fig. 4.5. Molecular structure of the osmium-formaldehyde complex [(CO)₂(PPh₃)₂Os(π-CH₂O) (Brown et al., 1979). Reproduced by permission of the American Chemical Society

species has parallels in molecular coordination chemistry. The first stable formaldehyde transition metal complex was prepared by Roper and coworkers (Brown et al., 1979), according to

$$(CO)_2(PPh_3)_3Os + CH_2O \rightarrow (CO)_2(PPh_3)_2Os(CH_2O) + PPh_3. \qquad (4.15)$$

The ligand exchange takes place at room temperature (benzene solution). The structure of the complex, as determined by X-ray single crystal analysis, is reproduced in Fig. 4.5. It shows the formaldehyde molecule π bonded (dihapto coordinated) to the metal. The strength of the π bond between metal and aldehyde CO is indicated by the considerable weakening of the C=O bond, as compared with the free formaldehyde (see Table 4.1). Several other formaldehyde complexes are included in the Table. The iron complex was synthesized from a μ-dinitrogen complex which reacts at room temperature with formaldehyde (in ether solution):

$$\begin{array}{ccc} \underset{OC'}{\overset{OC}{\diagdown}}\underset{\underset{L}{|}}{\overset{\overset{L}{|}}{Fe}}-N\equiv N-\underset{\underset{L}{|}}{\overset{\overset{L}{|}}{Fe}}\overset{\diagup CO}{\diagdown CO} & +\ CH_2O\ \longrightarrow & \underset{OC'}{\overset{OC}{\diagdown}}\underset{\underset{L}{|}}{\overset{\overset{L}{|}}{Fe}}\cdots\underset{O}{\overset{\overset{H\ H}{\diagdown\diagup}}{\overset{\|}{C}}}\ +\ N_2 \end{array}$$

(L = P(OCH₃)₃). The vanadium and zirconium complexes in Table 4.1 were found and investigated by Floriani and his school. The vanadium complex was obtained by interaction of vanadocene with paraformaldehyde. The zirconium-(μ-CH₂O) complex forms, interestingly, directly from a zirconium hydride complex and CO at room temperature and normal pressure (THF):

$$2(\eta^5\text{-}C_5H_5)_2Zr(H)(Cl) + CO \rightarrow [(\eta^5\text{-}C_5H_5)_2ZrCl]_2(\mu\text{-}CH_2O)$$

Table 4.1. Transition metal-formaldehyde complexes: stretching frequencies v_{CO} and C–O bond length of coordinated CH_2O.

Metal Complex	v_{CO} (cm^{-1})	C–O Bond length (pm)	Ref. [a]
CH_2O	1744	122.5	[1], [2]
$(PPh_3)(CO)_2Os(\pi\text{-}CH_2O)$	1017	159	[3]
$[(CH_3O)_3P]_2(CO)_2Fe(\pi\text{-}CH_2O)$	1220	132	[4]
$(\eta^5\text{-}C_5H_5)_2V(\pi\text{-}CH_2O)$	1160	135	[5]
$[(\eta^5\text{-}C_5H_5)ZrCl]_2(\mu\text{-}CH_2O)$	1015	143	[6]

[a] [1] Davis et al. (1967);
 [2] Wells (1962);
 [3] Brown et al. (1979);
 [4] Berke et al. (1981);
 [5] Gambarotta et al. (1982);
 [6] Gambarotta et al. (1983).

Electron diffraction analysis of the crystalline material shows that the formaldehyde molecule is π-bonded to one Zr center and σ-bonded, through the oxygen atom, to the other Zr unit:

$$Zr \cdots \begin{array}{c} O-Zr \\ \parallel \\ CH_2 \end{array}$$

Floriani (1983) assumes that CO insertion into a Zr–H bond of one Zr complex leads to a transient π-bonded formyl species which is then further reduced by a second hydride species.

The data in Table 4.1 show clearly a destabilization of the formaldehyde C=O bond on coordination, as expressed by the decrease in v_{CO} and the increase of the C–O bond length.

4.2.4 A Methoxy Complex

Bercaw and his coworkers found that treatment of the complex $(\eta^5\text{-}C_5Me_5)_2Zr(CO)_2$ with H_2 affords a methoxyhydrido complex in nearly quantitative yield:

$$(\eta^5\text{-}C_5Me_5)_2Zr(CO)_2 + 2\,H_2 \xrightarrow[110\,°C]{2\,atm} (\eta^5\text{-}C_5Me_5)Zr(H)(OCH_3) + CO. \qquad (4.16)$$

The detailed mechanism of this clean, stoichiometric reaction is not known, but the intermediate formation of a formyl ligand has been considered (Wolczanski and Bercaw, 1980, and references therein).

On the other hand, starting with $(\eta^5\text{-}C_5Me_5)ZrH_2$, and reacting it with CO, only minor amounts of the methoxy complex are observed, the main reaction product being the dimer

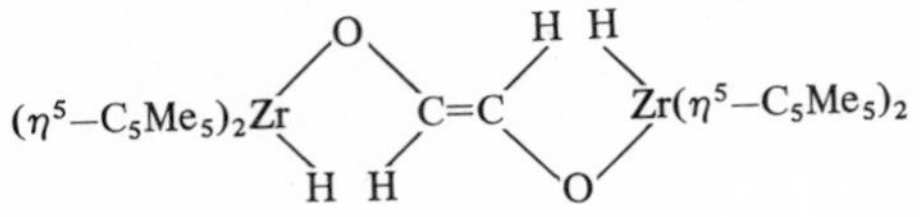

$$(\eta^5-C_5Me_5)_2Zr \underset{\underset{H\ H}{}}{\overset{\overset{O}{}}{\diagdown\diagup}} C=C \underset{\underset{O}{}}{\overset{\overset{H\ H}{}}{\diagup\diagdown}} Zr(\eta^5-C_5Me_5)_2$$

This is an amazing case of near stoichiometric carbon-carbon bond formation from CO under very mild conditions, at 1 atm and room temperature (Wolczanski and Bercaw, 1980).

4.2.5 Carbide Complexes

As mentioned in Sect. 3.4, surface metal-carbide species have also molecular models, in the form of transition metal carbide cluster compounds, whereby the carbidic carbon may procede from a CO molecule (see e.g. Fig. 3.9). However, in contrast to surface carbides which react easily with hydrogen to give mainly methane, most molecular carbide clusters are quite inactive, in particular those where the carbidic carbon is located in the center of the metal core of the cluster. Great hopes were attached to the peripheral carbide clusters, such as the one shown in Fig. 3.9, where the carbide carbon has a low coordination number, is sterically available, and the metal cluster shows coordination unsaturation (see e.g. Davis et al., 1981). This type of clusters had been considered not only as a model for the surface carbon reactions, but also as possible catalysts on their own. But although these compounds show indeed an interesting chemistry (Bradley et al., 1981; Muetterties, 1983; Holt et al., 1982; and references therein), in no case has the carbon been hydrogenated to CH_4 as yet. Protonation occurs on the metal bonds rather than on the carbidic carbon, with formation of a H bridge (Holt et al., 1982):

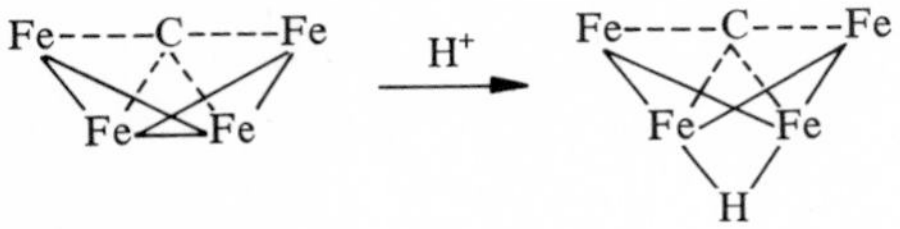

The reaction with hydrogen, in the presence of an oxidation agent, on the other hand, was suggested to lead to a dihapto bonded CH species containing also a bridging hydrogen (Davis et al., 1981; cf. Eq. 3.4):

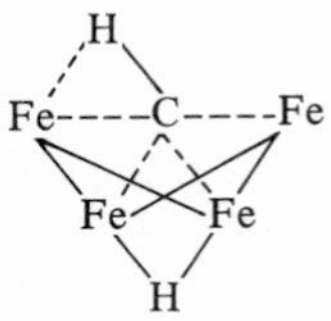

4.3 Conclusions

Convincing evidence is by now available for the interaction of CO and H_2, on metal surfaces as well as in molecular complexes, under mild conditions. Formyl complexes and π-coordinated formaldehyde appear to be among the most probable primary reaction products, as long as molecular (undissociated) carbon monoxide is involved. In many instances these early reaction intermediates (in particular formyl complexes) may not be thermodynamically stable, and the equilibria involved in their formation may lie far on the side of the dissociated components. But there appears to be no reason to deny the presence of such intermediates in kinetically significant amounts, available for further reaction (see, e.g. Sect. 7.2).

In those cases where carbon monoxide tends to dissociate on a metal surface, the resulting metal carbide will most probably react with hydrogen to give methane (and small amounts of higher hydrocarbons). In some cases this may be a major route to methane, in others a mere side reaction (cf. Sect. 7.1 and 9.5.2).

4.4 References

Audier M, Coulon M, Bonnetain L (1979) Carbon 17:391
Bahr HA, Bahr TH (1928) Ber Dtsch Chem Ges 61:2177
Basolo F, Pearson RG (1967) Mechanism of Inorganic Reactions. Wiley, New York, 2nd Ed.,
    p 555
Belmonte PA, Cloke FGN, Schrock RR (1983) J Amer Chem Soc 105:2643
Benziger JB, Madix RJ (1982) Surface Sci 115:279
Berke H, Bankhardt W, Huttner G, v. Seyerl J, Zsolnai L (1981) Chem Ber 114:2754
Bertolini JC, Imelik B (1979) Surface Sci 80:586
Bradley JS, Ansell GB, Leonowicz ME, Hill EW (1981) J Amer Chem Soc 103:4968
Brown KL, Clark GR, Headford CEL, Marsden K, Roper WR (1979) J Amer Chem Soc
    101:503
Byers BH, Brown TL (1977) J Organometal Chem 127:181
Cameron A, Smith VH, Baird MC (1983) Organometallics 2:465
Casey, CP, Neumann SM, Mark AA, McAlister DR (1980) Pure Appl Chem 52:625
Church MJ, Mays MJ, Simpson RNF, Stefanini FP (1970) J Chem Soc a 2909:3000
Collman JP, Winter SR (1973) J Amer Chem Soc 95:4089
Conrad H, Ertl G, Küppers J, Latta EE (1977) Proceed 6th Intern Congr Catalysis # 541.395
    IN
Craig JH (1982) Appl Surface Sci 10:315
Davis JH, Beno MA, Williams JM, Zimmie J, Tachikawa M, Muetterties EL (1981) Proc
    Natl Acad Sci USA 78:668
Davis RE, Grosse D, Ohno A (1967) Tetrahedron 23:1029
Deluzarche A, Hindermann JP, Kieffer R (1978) Tetrahedron Lett 31:2787
Dombek BD, Harrison AM (1983) J Amer Chem Soc 105:2486
Fagan PJ, Moloy KG, Marks TJ (1981) J Amer Chem Soc 103:6959
Fischer JR, Mills AJ, Sumner S, Brown MP, Thomson MA, Puddephatt RJ, Frew AA,
    Manojlovic-Muir L, Muir KW (1982) Organometallics 1:1421
Floriani C (1983) Pure Appl Chem 55:1

Galwey AK (1962) J Catal 1:227
Gambarotta S, Floriani C, Chiesi-Villa A, Guastini C (1982) J Amer Chem Soc 104:2020
Gambarotta S, Floriani C, Chiesi-Villa A, Guastini C (1983) J Amer Chem Soc 105:1690
Gladysz JA (1982) Adv Organometal Chem 20:1
Goodman DW, Kelley RD, Madey TE, Yates JT (1980) J Catal 63:226
Holt EM, Whitmire KH, Shriver DF (1982) J Amer Chem Soc 104:5621
Kieffer EPh, van der Baan HS (1982) Appl Catal 3:245
Kraemer K, Menzel D (1975) Ber Bunsenges Phys Chem 79:649
McCarty JG, Wise H (1979) J Catal 57:406
Muetterties EL (ed) (1972) Transition Metal Hydrides. Marcel Dekker, New York
Muetterties EL, Krause MJ (1983) Angew Chem 95:135; Angew Chem Int Ed Engl 22:135
Ogoshi H, Setsune J, Omura T, Yoshida Z (1975) J Amer Chem Soc 97:6461
Peebles DE, Schreifels JA, White JM (1982) Surface Sci 116:117
Primet M, Sheppard N (1976) J Catal 41:258
Rabo JA, Risch AP, Poutsma ML (1978) J Catal 53:295
Rostrupp-Nielsen J, Trimm DL (1977) J Catal 48:155
Saussey J, Lavalley J-C, Lamotte J, Rais T (1982) J Chem Soc Chem Commun 278
Thorn DL (1982) Organometallics 1:197
Vannice MA (1976) Catal Rev-Sci Eng 14:153
Vidal JL, Walker WE (1981) Inorg Chem 20:249
Vlasenko VM, Kukhar LA, Rusov MT, Samchenko NP (1964) Kinet Catal USSR 5:301
Wayland BB, Duttaahmed A, Woods BA (1983 J Chem Soc Chem Commun 142
Wells AF (1962) Structural Inorganic Chemistry. 3rd Ed., Oxford Press, p 717
Wolczanski PT, Bercaw JE (1980) Acc Chem Res 13:121
Wong WK, Tam W, Strouse CE, Gladysz JA (1979) J Chem Soc Chem Commun 530
Worley SD, Mattson GA, Caudill R (1983) J Phys Chem 87:1671

5 Key Reactions in Catalysis

This Section reviews briefly those classes of reactions which play an important role in the catalytic hydrogenation of carbon monoxide. Some of the reactions have already been mentioned in earlier Chapters (e.g. oxidative addition of hydrogen in Sect. 2.1, ligand exchange in Section 4.2.2), but it is considered convenient to group the different types of reactions together here for easy reference in later Chapters. It should be noted that these "key reactions" are ubiquitous in catalysis with transition metals in general (certainly in homogeneous, and most probably also in heterogeneous catalysis), but only those aspects relevant to the hydrogenation of carbon monoxide will be discussed. For a broader treatment, the reader is referred to the specialized literature (e.g. Henrici-Olivé and Olivé, 1977; Collman and Hegedus, 1980; Cotton and Wilkinson, 1980).

5.1 Oxidative Addition

The term "oxidative addition" is used for the addition of neutral XY molecules, such as H_2 or alkyl halides, to transition metal species having coordination sites available. Formally, the XY molecule is reductively dissociated to give two anionic ligands X^- and Y^-, and the metal is simultaneously oxidized. The prototype of this rather widespread class of reactions was detected by Vaska and DiLuzio (1962), who added H_2 to a square planar iridium (I) complex (cf. Eq. 2.1):

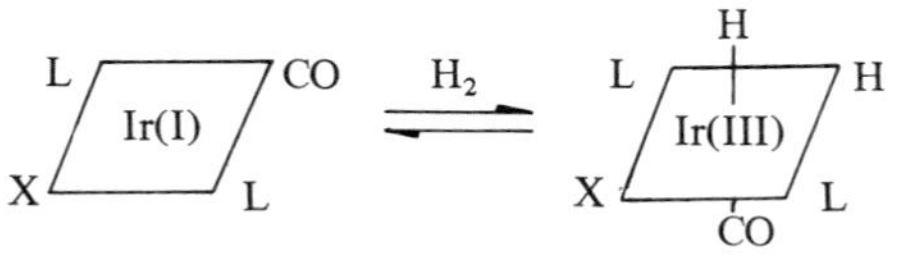

$$
\begin{array}{c}
\text{L} \diagup \!\!\!\!\!\!\diagdown \text{CO} \\
\quad \fbox{Ir(I)} \\
\text{X} \diagdown \!\!\!\!\!\!\diagup \text{L}
\end{array}
\quad \xrightleftharpoons{H_2} \quad
\begin{array}{c}
\quad\quad \text{H} \\
\text{L} \diagup \!\!\!\!\!\!\diagdown \text{H} \\
\quad \fbox{Ir(III)} \\
\text{X} \diagdown \!\!\!\!\!\!\diagup \text{L} \\
\quad \text{CO}
\end{array}
\tag{5.1}
$$

(X = Cl, Br, I; L = PPh$_3$). Actually, the newly formed metal-hydrogen bonds are covalent σ-bonds. However, according to a generally accepted convention, the "oxidation number" of a transition metal in a complex is defined as the charge remaining on the metal when each electron pair shared by the metal and a ligand is formally assigned to the ligand. The only shared electron pair in the square planar complex (Eq. 5.1) is that of the $Ir-X$ bond; the other three bonds are coordinative bonds where the ligand contributes both electrons. Hence, the iridium in the square planar complex has the oxidation number I. In the octahedral complex resulting from the H_2 addition, there are three shared electron pairs, one of the $Ir-X$, and two of the $Ir-H$ bonds; hence the oxidation number of the metal is III. This example shows that oxidation numbers, though providing a useful formalism, are not related to any measurable physical property of the metal species, in particular they cannot inform about the actual charge on the metal.

As to the availability of free coordination sites — evidently a prerequisite for oxidative addition — it is useful to remember that complexes in which the metal has an 18 electron ("noble gas") configuration tend to be particularly stable; they are said to be coordinatively saturated. The "18 electron rule" may be easily interpreted in terms of the molecular orbital theory of transition metal complexes (see e.g. Henrici-Olivé and Olivé, 1977). The noble gas configuration is given if the sum of the d electrons of the metal in its particular valency state, two electrons for each covalently or coordinatively bonded ligand, and x electrons for each η^x-bonded ligand, is eighteen. For ionic species the overall charge has to be taken into account. For instance, the following stable species have an 18 electron configuration:

$$Cr(CO)_6; \quad (\eta^5\text{-}C_5H_5)_2Fe; \quad RhH(CO)(PPh_3)_3; \quad [Co(CN)_6]^{3-}.$$

From the above follows a maximum number of ligands for each d^n metal configuration (maximum coordination number CN_{max}). For a neutral metal complex, without oligohapto ligands, it is given by

$$CN_{max} = \frac{18-n}{2}. \tag{5.2}$$

Complexes short of this coordination number are often prone to oxidative addition reactions. This is particularly true for square planar d^8 complexes, such as the Ir(I) complex in Eq. (5.1), which has an electron count of 16 and, according to Eq. (5.2), $CN \leq CN_{max} = 5$ for Ir(I). The oxidative addition of H_2 leads to an electron count of 18 and to the maximum allowed number of 6 ligands for Ir(III) (d^6). Complexes with 18 electrons cannot undergo oxidative addition unless previous expulsion of a ligand occurs.

Oxidative addition of H_2 to binuclear cobalt carbonyls such as $Co_2(CO)_8$ (cf. Eq. 2.2) and $Co_2(CO)_6(PR_3)_2$ leads to mononuclear monohydrides e.g.:

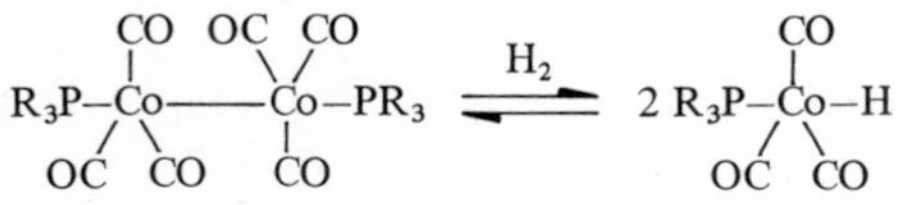

$$\tag{5.3}$$

The binuclear d^9 species has an electron count of 18 for each Co center (one electron for the metal-metal bond). Breaking of the Co$-$Co bond results in a (hypothetical) 17 electron species which can accept only one hydrogen ligand. The mononuclear d^8 species has its maximum coordination number of 5. (The actual reaction mechanism may involve the dissociation of one ligand in a preequilibrium; Collman and Hegedus, 1980.)

Metal atoms located in the outermost layer of clean metal surfaces are necessarily coordinatively unsaturated and should be available for oxidative addition of hydrogen. Whether this action leads primarily to dihydrido-surface species related to that in Eq. (5.1), or to monohydrido species as in Eq. (5.3) is not known at present. It may depend on the situation of the metal atom (atoms at steps or crystalline irregularities have certainly more coordination sites available than those in the middle of a plain surface), on the metal itself (the lower the d-population, the higher the maximum coordination number, see Eq. (5.2); hence all other conditions comparable, the tendency to form di-hydrides by mononuclear oxidative addition should increase in the series Ni < Co < Fe and Pd < Rh < Ru), and of course on other possible ligands (in particular CO) present on the surface.

Vaska and Werneke (1971) have pointed to the analogy between the oxidative addition of H_2 to molecular complexes and that to metal surfaces, based on the similarity of activation and thermodynamic data. Thus, (the activation enthalpy for the addition of H_2 to the square planar "Vaska complex" (Eq. 5.1), although somewhat dependent on the ligand X ($\Delta H^{\ddagger} \simeq 25-50$ kJ/mol) is in the same range as those for metal surfaces (8$-$46 kJ/mol); also the overall heat for the H_2 addition reaction, $-\Delta H^0$, is not very different for the iridium complexes (50$-$80 kJ/mol) and on iridium surfaces ($\simeq 108$ kJ/mol).

The notation oxidative addition does not imply any particular mechanism, such as concerted, stepwise, or radical. In the case of the H_2 addition to *molecular transition metal complexes*, the reaction is generally believed to be a concerted three-center process with simultaneous bond breaking and bond making, so that the energetically favorable formation of the two metal-hydrogen bonds contributes to the dissociation of the H$-$H bond (435 kJ/mol). Theoretical considerations (Noell and Hay, 1982) concerning the addition of H_2 to PtL_2 (L$=$P(CH$_3$)$_3$ or PH$_3$) led to the suggestion that the hydrogen molecule approaches the metal center "side-on", and that the major electronic contribution in the transition state is the donation of d-electron density from the metal $5\,d_{xy}$ orbital into the antibonding σ^* hydrogen orbital, thus weakening the H$-$H bond.

$$L_nM + H_2 \rightleftarrows \left[L_nM \overset{..H}{\underset{..H}{\vdots}} \right] \rightarrow L_nM \overset{H}{\underset{H}{<}} . \tag{5.4}$$

The reaction should lead to cis-dihydride complexes, and in fact this is generally its outcome. However, even if trans-dihydrides are found, this does not necessarily indicate a different mechanism, since isomerization of the initial product may occur. The final product will be the isomer (or isomer mixture) thermodynamically most stable under the pertaining conditions.

On the other hand, Shustorovich et al. (1983) using a different theoretical model for the oxidative addition of H_2 (and other molecules) on *metal surfaces*, suggested that the approach of the H_2 molecule is "end-on", and that the "side-on" position corresponds rather to an excited state preceding the reaction proper. Of the two configurations for the excited state (*a* and *b* below) *a* was found to be more favorable. Intuitively, this would be expected to lead to monohydridic species:

$$
\begin{array}{cc}
\text{H}\cdots\text{H} & \text{H}\cdots\text{H} \\
\vdots\quad\vdots & \ddots\ddots \\
-\text{M}-\text{M}- & -\text{M}-\text{M}-\text{M}- \\
a & b
\end{array}
$$

However, these considerations were explicitly limited to atomically flat clean metal surfaces, and it was hinted that the situation may be different for atoms at topographic irregularities such as steps and kinks. Since more coordination sites are available at such locations, structure *b* may become more favorable, and lead to mononuclear addition of the hydrogen molecule.

5.2 Reductive Elimination

Frequently (but not generally) oxidative addition is reversible. The term "reductive elimination" is used for the reverse reaction. Intra-, as well as internuclear eliminations have been recognized:

$$
\begin{array}{c}
\text{Y} \\
| \\
\text{L}_n-\text{M}-\text{X} \rightarrow \text{L}_n\text{M} + \text{XY}
\end{array}
\tag{5.5}
$$

$$
\begin{array}{cc}
\text{Y} & \text{Y} \qquad\quad \text{Y}\ \text{X} \\
| & | \qquad\qquad\quad |\ \ | \\
\text{L}_n-\text{M}-\text{X} + \text{L}_n-\text{M}-\text{X} \rightarrow \text{L}_n-\text{M}-\text{M}-\text{L}_n + \text{XY}.
\end{array}
\tag{5.6}
$$

In catalysis, it often occurs that one of the oxidatively added ligands undergoes further reactions (e.g. insertion of an alkene or CO molecule, see Sect. 5.3) before reductive elimination removes the reaction product, regenerating the catalyst.

Reductive eliminations can involve (among others) an alkyl group and a hydride ligand giving alkane, an acyl ligand and hydride resulting in an aldehyde, or two alkyl groups forming an alkane. Some examples are discussed below. Often, it proves difficult to distinguish between intra- and intermolecular reductive eliminations. However, isotopic labeling can help to decide between the two possibilities. The complex under investigation is mixed with its isotopic isomer having the leaving groups perdeuterated. The absence of crossover product indicates clearly intramolecular reductive elimination.

5.2.1 Intramolecular Reductive Elimination

The oxidative addition of H_2 (e.g. Eq. 5.1) is commonly reversible; the reductive elimination can often be attained by evacuation at 25 °C (Cotton and Wilkinson, 1980). The displacement of hydride ligands by CO which has been observed in several occasions (Brown et al., 1980 and references therein; see also Eqs. 4.3 and 4.4) is to be considered as a reductive elimination because covalent ligands are displaced by coordinatively bonded ligands, i.e. the oxidation number of the metal is reduced; e.g.

$$L_nM\overset{-H}{\underset{-H}{}} \underset{H_2}{\overset{CO}{\rightleftharpoons}} L_nM\overset{-CO}{\underset{-CO}{}}.$$

The clean intramolecular reductive elimination of methane from the square complex cis-PtH(CH$_3$)(PPh$_3$)$_2$ has been studied by Halpern and his school (Abis et al., 1978). The complex is stable at -50 °C, and decomposes according to Eq. (5.7) at a convenient, measurable rate at -25 °C.

$$\text{cis-PtH(CH}_3\text{)(PPh}_3\text{)} \xrightarrow{k_1} \text{Pt(PPh}_3\text{)}_2 + \text{CH}_4. \tag{5.7}$$

The intramolecular character of the process was confirmed by means of the above mentioned crossover experiment. Kinetic measurements yielded a first order rate law; the rate constant k_1 ($4.5 \times 10^{-4} \, s^{-1}$ at -25 °C) was unaffected if excess ligand PPh$_3$ was added. These observations were interpreted in terms of a mechanistic scheme encompassing the rate determining step (5.7), followed by

$$\text{Pt(PPh}_3\text{)}_2 + \text{PPh}_3 \xrightarrow{\text{fast}} \text{Pt(PPh}_3\text{)}_3.$$

In the absence of excess ligand, Pt(PPh$_3$)$_2$ disproportionated to Pt(PPh$_3$)$_3$ and Pt.

Gillie and Stille (1980), investigated the reductive elimination of ethane from cis- and trans-$L_2Pd(CH_3)_2$ (L = PPh$_3$) in polar solvents. By means of labeling experiments it was shown that the reaction is intramolecular (cf. Eq. 5.5) in both cases. However, no reductive elimination took place with the trans isomer in non-polar solvents even at elevated temperatures. It was found that a rapid isomerization trans-cis preceded the reductive elimination in polar solvents. (A five coordinate transition state where a solvent molecule occupies one coordination site was suggested.) These experiments indicate the necessity of a cis-position of the leaving ligands in intramolecular reductive elimina-tions. In contrast to the Pt case discussed above (Eq. 5.7), the reaction rate was reduced by the presence of excess ligand, indicating that a phosphine ligand needs to be dissociated before the reductive elimination can take place. This appears somewhat enigmatic, since no vacant coordination site is required for reductive elimination. However, the experimental result was rationalized by Hoffmann (1981) by theoretically showing that the reductive elimination is easier from a three coordinate than from a four coordinate d^8 complex, the latter having an energy barrier at least twice as high as the former. Hence, the reaction sequence is, most probably, given by Eq. (5.8).

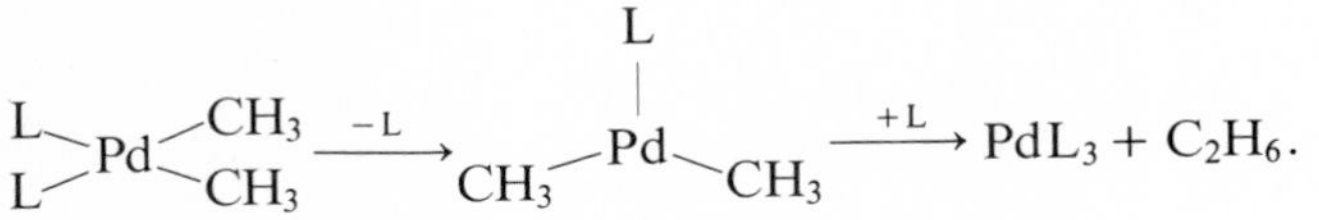

$$\underset{L}{\overset{L}{>}}Pd\overset{CH_3}{\underset{CH_3}{<}} \xrightarrow{-L} CH_3-\underset{\overset{|}{L}}{Pd}-CH_3 \xrightarrow{+L} PdL_3 + C_2H_6. \qquad (5.8)$$

The reductive elimination of acetone from the octahedral d^6 complex cis-RhH(CH$_2$COCH$_3$)L$_3$Cl (L = PMe$_3$) has been found by Milstein (1982) to be intramolecular by the same labeling method. Hoffmann (1981) predicted reductive elimination from octahedral d^6 complexes also to be of the dissociative type, and in fact the reaction rate was reduced by the presence of excess ligand L, indicating a rate determining loss of L prior to the reductive elimination:

$$
\begin{aligned}
\text{cis-RhHRL}_3\text{Cl} &\rightleftarrows \text{RhHRL}_2\text{Cl} + \text{L} \\
\text{RhHRL}_2\text{Cl} &\rightarrow \text{RhL}_2\text{Cl} + \text{RH} \qquad\qquad (5.9) \\
\text{RhL}_2\text{Cl} + \text{L} &\rightarrow \text{RhL}_3\text{Cl} .
\end{aligned}
$$

5.2.2 Intermolecular Reductive Elimination

Intermolecular reductive elimination is a facile reaction, if at least one of the ligands to be eliminated is a hydride, and if the complex bearing the other ligand to be eliminated has a vacant coordination site in cis position (Norton,

1979). The unique ability of hydride ligands to bridge pairs of transition metal atoms is assumed to be a major factor in the process, and the reaction is believed to be initiated by such bridge formation:

$$L_nM-H + L_{n-1}M-R \rightarrow L_nM-H-ML_{n-1}. \qquad (5.10)$$

(A)   (B)   R

The vacant coordination site on complex B may be present from the beginning; more frequently a site is vacated in the course of the process. Some examples follow.

The reductive elimination of H_2 from $Os(CO)_4H_2$ is intermolecular, as demonstrated by labeling crossover. Nevertheless, the rate is first order with regard to the starting compound. This suggests a rate determining step involving carbon monoxide dissociation, followed by fast binuclear elimination of hydrogen (including the bridge formation, Eq. 5.10), and recoordination of CO (Norton, 1979):

$$Os(CO)_4H_2 \xrightarrow{\text{slow}} Os(CO)_3H_2 + CO$$

$$Os(CO)_3H_2 + Os(CO)_4H_2 \xrightarrow{\text{rapid}} H_2Os_2(CO)_7 + H_2 \qquad (5.11)$$

$$H_2Os_2(CO)_7 + CO \xrightarrow{\text{rapid}} H_2Os_2(CO)_8.$$

In the following example, the coordination site is vacated by a migratory insertion reaction (cf. Sect. 5.3). Jones and Bergman (1979) discovered that $CpMoH(CO)_3$ and $CpMoR(CO)_3$ ($Cp = \eta^5-C_5H_5$, $R = CH_3$ or C_2H_5) undergo a clean and quantitative reaction which leads to the aldehyde RCHO and dimeric molybdenum species, at temperatures between 25 and 50 °C. In contrast to the preceding example, these reactions give second-order kinetics. The mechanism is assumed to include a rapid preequilibrium

$$R-Mo(CO)_3Cp \rightleftarrows R-\overset{\displaystyle O}{\overset{\|}{C}}-Mo(CO)_2Cp. \qquad (5.12)$$

By inserting one of the CO ligands into the Mo−R bond, an acyl ligand is formed, and a vacancy is created. This is followed by the rate determining bimolecular elimination reaction:

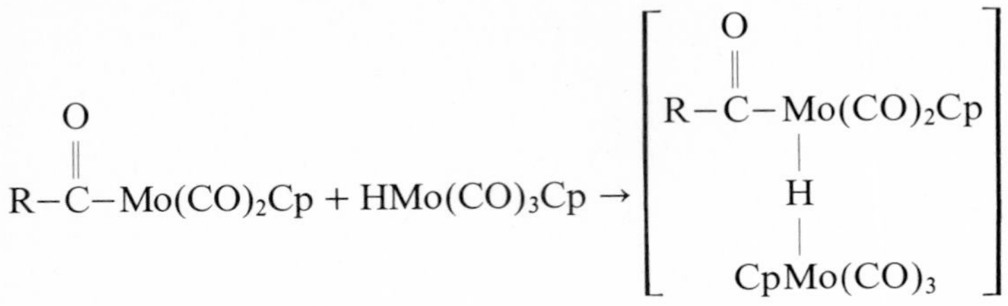

$$
\begin{array}{c}
O \\
\parallel \\
R-C-Mo(CO)_2Cp + HMo(CO)_3Cp
\end{array}
\rightarrow
\left[
\begin{array}{c}
O \\
\parallel \\
R-C-Mo(CO)_2Cp \\
| \\
H \\
| \\
CpMo(CO)_3
\end{array}
\right]
$$

$$
\rightarrow RCHO + Cp_2Mo_2(CO)_4 + Cp_2Mo_2(CO)_6. \qquad (5.13)
$$

It may be noted that it is assumed that this relatively facile reaction pattern makes hydridoalkyl metal carbonyls inherently unstable (Norton, 1979).

For the sake of completeness it is mentioned that ligands R giving rise to very stable radicals, such as a p-substituted benzyl ligand, may leave the complex via homolytic cleavage, with a subsequent free radical intermolecular reductive elimination pathway (Halpern, 1982); however special conditions concerning the solvent, the other ligands in the complex, as well as the type of p-substitution of the benzyl ligand have to be met in order this to happen (Nappa et al., 1982).

In our present context it is important to remember that both, *intra- and intermolecular reductive elimination do exist*. It should, however, be born in mind that the intermolecular reaction which, according to present knowledge, always involves the transfer of a hydride ligand from one complex to a ligand R (R=H, alkyl, acyl, etc.) on the second unit, requires mobility of the hydride-bearing complex, so that it can approach the other unit from the right side, and with the right angle in order to form the μ-H intermediate (see Eqs. 5.10 and 5.13). It appears evident that this mobility is warranted only in solution, and it has been suggested (Henrici-Olivé and Olivé, 1977/78) that this might be one of the major mechanistic differences between homogeneous hydrogenations of CO (such as hydroformylation, and the synthesis of glycol and other polyols, Chap. 10) on the one side, and heterogeneous reactions (such as methanation, Chap. 7, and the Fischer-Tropsch synthesis, Chap. 9) on the other.

5.3 Migratory Insertion Reactions

This class of reactions involves the combination of a saturated σ-bonded ligand R with an unsaturated ligand coordinatively bonded to the same metal center, forming one new ligand. In our context the most important insertion reaction is that of carbon monoxide

$$
\begin{array}{cc}
R & L' \\
| & | \\
L_n M-CO \rightleftharpoons L_n M-C-R \\
& \| \\
& O
\end{array}
\qquad (5.14)
$$

(R = alkyl or H) which leads to the formation of an acetyl or a formyl ligand. Carbon monoxide insertion forms part of all homogeneous, and probably most heterogeneous catalytic CO hydrogenation processes. The essential characteristics of the reaction will be discussed mainly in this Section. The other important insertion reaction, that of olefins:

$$
\begin{array}{c}
R \quad \backslash / \\
| \quad C \\
L_n M \cdots \| \rightarrow L_n M - C - C - R \\
C \\
/ \backslash
\end{array}
\qquad (5.15)
$$

will only be touched briefly, since it plays only a minor role in CO hydrogenation (as a secondary reaction of olefins formed in the Fischer-Tropsch synthesis).

5.3.1 CO Insertion into Metal-Alkyl Bonds

Carbon monoxide insertions into metal-alkyl σ bonds have been reported for almost all transition metals (Calderazzo, 1977, and references therein). One of the best known examples is methylmanganesepentacarbonyl which, in the presence of ligands L (such as CO or phosphine), forms the corresponding acyl complex according to Eq. (5.14). The mechanism has been studied in considerable detail by Calderazzo (1977).

Using ^{13}CO as an incoming ligand L' (Eq. 5.14), it was shown by means of IR spectroscopy, that one of the CO ligands present in the complex ends up as part of the acetyl ligand, whereas the incoming ligand ^{13}CO is located in cis position:

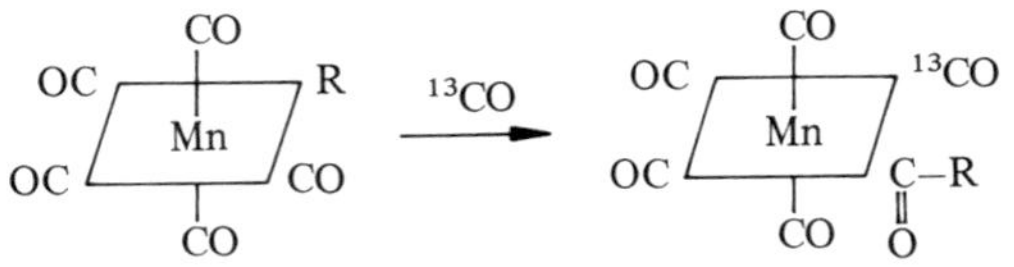

$$
\qquad (5.16)
$$

Kinetic measurements of the reaction at varying carbon monoxide pressure indicated that the rate of formation of the acetyl complex, at each applied CO pressure, is pseudo-first order with regard to the concentration of the starting methyl complex:

$$\text{rate} = k_{obs} \, [CH_3Mn(CO)_5] \tag{5.17}$$

whereby k_{obs} depends on [CO] at low CO pressure and becomes independent at high pressure. This is best interpreted in terms of a two step process:

$$CH_3-Mn(CO)_5 \underset{k_{-1}}{\overset{k_1}{\rightleftharpoons}} [CH_3\overset{\|}{\underset{O}{C}}-Mn(CO)_4] \underset{k_{-2(-L)}}{\overset{k_2(+L)}{\rightleftharpoons}} CH_3\overset{\|}{\underset{O}{C}}-Mn(CO)_4L \tag{5.18}$$

(L = CO). By applying the steady state approximation, and under the reasonable assumption that k_{-2} is comparatively small, the following kinetic equations apply (C = starting methyl complex, I = five coordinate intermediate, P = product acetyl complex):

$$\frac{d[I]}{dt} = k_1[C] - (k_1 + k_2) \, [I] = 0$$

$$[I] \quad = k_1[C]/k_{-1} + k_2$$

$$\text{rate} = k_2[I] \, [L] \quad = \frac{k_1 \, k_2[C] \, [L]}{k_{-1} + k_2[L]} \, .$$

Hence, k_{obs} (Eq. 5.17) becomes:

$$k_{obs} = \frac{k_1 \, k_2[L]}{k_{-1} + k_2[L]}$$

which provides for the CO pressure dependence mentioned above (first order in [L] at very small [L] when $k_2[L] \ll k_{-1}$; zero order at large [L] when $k_2[L] \gg k_{-1}$; mixed order at intermediate [L]).

Carbon monoxide insertion is generally assumed to proceed by a concerted reaction path. Concerted reactions are characterized by relatively low activation energies; they are usually lower than the bond dissociation energies of the weakest bonds involved, indicative of concerted bond breaking and bond making. Activation entropies are either very small or negative, indicative of the restriction of motion resulting from the formation of a cyclic transition state (O'Neal and Benson, 1967). Activation parameters have been reported by Calderazzo and Cotton (1962a) for the insertion of CO according to Eq. (5.16): $E_a = 61.9$ kJ/mol; $\Delta H^* = 59.4$ kJ/mol; $\Delta F^* = 86.2$ kJ/mol; $\Delta S^* = -88.3$ kJ/mol. The bonds to be broken are the metal$-$C (methyl) and the metal$-$C (CO) bonds. Metal-alkyl σ bonds are generally in the range $160-350$ kJ/mol (Connor, 1977); in the particular case of $CH_3Mn(CO)_5$ an extremely low value of 125 kJ/mol has been reported (Lalage et al., 1974), but it is still high compared to the observed activation energy. Metal$-$CO bond dissociation energies for first row metals are in the neighborhood of 100 kJ/mol (see Table 3.3). Thus, the overall activation energy is considerably lower than the dissociation energy of each of the bonds involved.

An interesting aspect of the migratory insertion of CO into a metal-alkyl bond should be mentioned: it appears that the ease of the reaction depends strongly on the donor ability of the alkyl ligand. Thus the equilibrium constant for the insertion reaction

$$MnR(CO)_5 + CO \rightleftarrows Mn(CR)(CO)_5$$
$$\overset{\|}{\underset{O}{}}$$

at room temperature has values of 3300, 360 and 66 l/mol for R = ethyl, methyl, and phenyl, respectively (Calderazzo, 1977). The ratio of the rates of the forward reaction for R=ethyl or methyl is 13:1 (Calderazzo and Cotton, 1962b). Also in the case of a square planar platinum complex, cis-RPt(CO)LX (X = halide, L = tertiary phosphine), insertion is favored for R=C_2H_5 as compared to R=CH_3, and for a series of m- or p-substituted phenyl phosphine, the ease of inserting the CO ligand increases with increasing electron donating power of the phenyl ligand as indicated by the Hammet σ constants (Cross and Gemmill, 1981).

5.3.2 Alkyl Migration versus CO Migration

Some discussion centered upon the question whether insertion reactions such as (5.16) take place at the site of the σ bonded ligand, or whether this ligand migrates to the site of the coordinated substrate molecule:

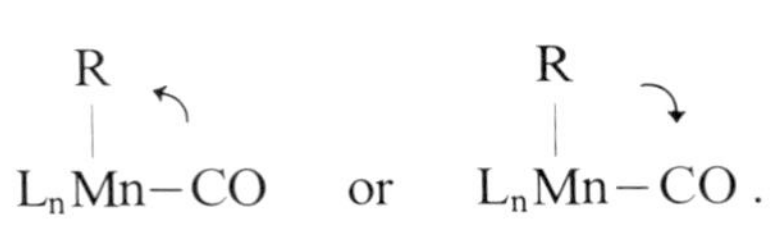

$$\overset{R}{\underset{L_nMn-CO}{|}} \quad \text{or} \quad \overset{R}{\underset{L_nMn-CO}{|}}.$$

For the case of the methylmanganesepentacarbonyl (Eq. 5.16) the question has been settled by Noack and Calderazzo (1967) in an elegant way, studying

the reverse of (5.16), i.e. the thermal decarbonylation of the acetyl complex
cis-CH$_3$CMn^{13}CO(CO)$_4$. The reaction products were carefully analyzed (IR)

$$\overset{\|}{O}$$

with regard to their stereochemistry at the metal. The resulting isomer
distribution (cis- and trans-CH$_3$Mn^{13}CO(CO)$_4$ and unlabeled CH$_3$Mn(CO)$_5$)
could only be explained by cis-migration of the methyl group to a site vacated
by a leaving CO ligand. Based on the principle of microscopic reversibility,
it was assumed that the forward reaction (Eq. 5.16) takes place by methyl
migration to a cis-CO ligand rather than by CO insertion at the site of the
ligand R.

These results concerning the CH$_3$Mn(CO)$_5$ complex have been confirmed
by Flood et al. (1981) who used ^{13}C NMR as their analytical tool. These
authors tackled also the question of the geometry of the five-coordinate
intermediate, for which trigonal bipyramidal as well as square pyramidal
structures had previously been suggested. Their ^{13}C NMR data indicate a
square pyramidal structure, with an equatorial acyl ligand, for the intermediate:

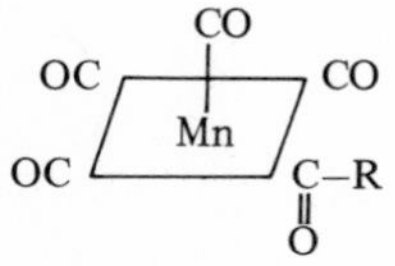

Such detailed studies are not available at the present for other complexes.
However, by analogy it is assumed that insertion reactions usually take the
indicated course, and the term "migratory insertion" has become customary
for the whole class of insertion reactions.

It may be noted that in one particular Rh complex CO insertion led to a
five coordinated acyl product which was isolated and structurally characterized by X-ray diffration (Cheng et al., 1977); the compound had the acyl
group in the axial position of a square pyramidal complex

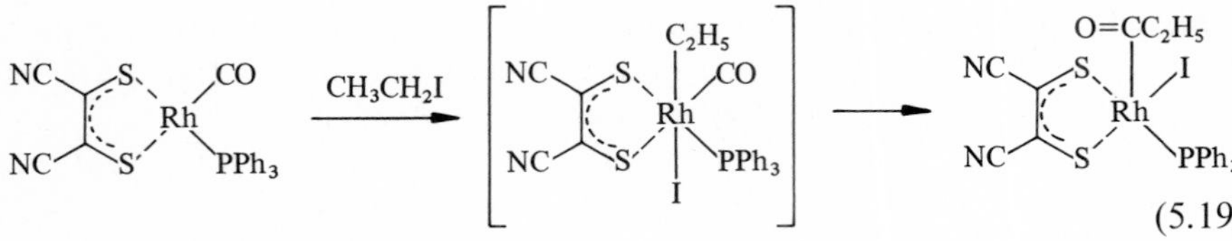

(5.19)

This result appears to imply CO insertion at the site of the alkyl group rather
than alkyl migration. However, since it cannot be excluded that an initial
product with an equatorial acyl ligand has subsequently rearranged to the
more stable axial product, the question has to remain open in this case.

Sakaki et al. (1983) tried to solve the problem theoretically, for the particular case of the two isomers of the square planar complex cis-$CH_3Pt(CO)F(PH_3)$, with the phosphin ligand in cis- or trans-position to the methyl group. An ab initio Hartree-Fock MO calculation led them to the result that for both complexes methyl migration is an easy reaction path, whereas CO migration to the site of the methyl ligand has too high a potential barrier to be a probable route. However, a mutual approach of the CH_3 and CO ligands, with simultaneous opening of the F–Pt–P angle, was found theoretically feasible also, at least in one of the isomers. This would correspond to a trigonal bipyramidal structure in the case of octahedral complexes, which then might rearrange to a more stable square pyramidal intermediate, as that deduced by Flood et al., from ^{13}C NMR data (vide supra).

To a certain degree it appears dangerous to generalize this type of experimental or theoretical treatment of any particular complex, since the detailed mechanism may depend not only on the ligands and the symmetry of the complex, but also on possible solvent intervention (see e.g. Wax and Bergman, 1981, and references therein). Therefore, unless otherwise stated, we shall use the term "migratory insertion", or simply "insertion" to merely describe the outcome of the reaction, without attributing any detailed mechanistic significance to it.

5.3.3 Promotion of the Migratory CO Insertion

Magnuson et al. (1983) reported a rapid, redox catalyzed migratory insertion with the complex $(\eta^5\text{-}C_5H_5)(PPh_3)(CO)Fe(CH_3)$. The formally Fe(II) complex reacts very sluggishly to give the corresponding acetyl complex:

$$Cp(PPh_3)(CO)FeCH_3 + CO \rightarrow Cp(PPh_3)CO\underset{\overset{\|}{O}}{Fe}CCH_3 \qquad (5.20)$$

For example in 5 days no perceptible ($< 5\%$) conversion could be observed (1 atm of CO, 0 °C, CH_2Cl_2). In striking contrast the addition of a few percent of an oxidant (ferricinium tetrafluoroborate or silver tetrafluoroborate) caused the conversion to the acetyl complex to be complete in $1-2$ min. This amazing redox catalysis is rationalized in terms of the following Scheme:

$$L_2Fe(CH_3)(CO) \xrightarrow{\text{ox}} [L_2Fe(CH_3)(CO)]^+$$

$$[L_2Fe(CH_3)(CO)]^+ + CO \rightarrow [L_2Fe(C\underset{\overset{\|}{O}}{C}H_3)(CO)]^+$$

$$[L_2Fe(C\underset{\overset{\|}{O}}{C}H_3)(CO)]^+ + L_2Fe(CH_3)(CO)$$

$$\rightarrow L_2FeC\underset{\overset{\|}{O}}{C}H_3(CO) + [L_2Fe(CH_3)(CO)]^+ .$$

The enhanced rate is attributed to a decrease of the $d\pi$ donor ability of the metal on going from Fe(II) to Fe(III). The presence of the formally Fe(III) complex is demonstrated by cyclic voltametry. A rapid electron transfer is assumed to restore the cationic methyl Fe(III) species, which needs to be present only in small amounts.

A strong influence of the gegenion of an anionic iron complex on the rate of insertion of CO was described by Collman et al. (1978) for the reaction:

$$[RFe(CO)_4]^- X^+ + PPh_3 \rightarrow [R\overset{\underset{\|}{O}}{C}Fe(CO)_3(PPh_3)]^- X^+ \tag{5.21}$$

(R = n-alkyl). In THF at 25 °C, the cations Li^+ and Na^+ cause the insertion reaction to occur 2–3 orders of magnitude faster than the rate observed if the cation is trapped in a crown ether, or if the bulky cation $(Ph_3P)_2N^+$ is used instead. IR and NMR evidence indicates that the interaction with the cations occurs via the electron rich oxygen of one of the carbonyl groups in the starting complex, and with the acyl oxygen in the product complex. It is assumed that the rate of alkyl migration is so dramatically increased by the presence of electron acceptors like Li^+ and Na^+ because they are able to stabilize the coordinatively unsaturated intermediate by forming a tight ion pair

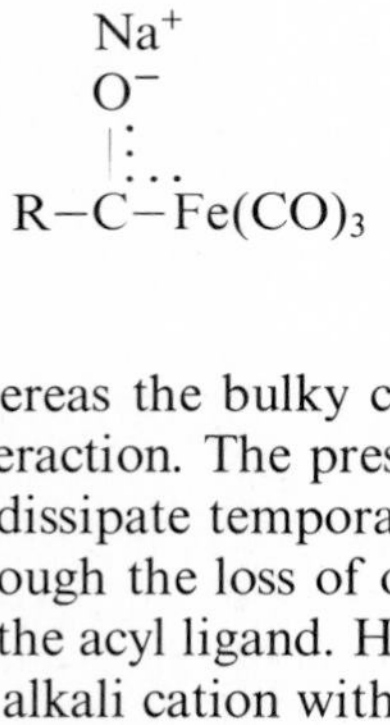

$$\begin{array}{c} Na^+ \\ O^- \\ \vdots \\ R-C-Fe(CO)_3 \end{array} \tag{5.22}$$

whereas the bulky cations cannot approach the oxygen close enough for such interaction. The presence of the electron acceptor cations Na^+ or Li^+ may help to dissipate temporarily electron density from the metal, released to the latter through the loss of one π-acceptor ligand (CO) in the course of the formation of the acyl ligand. However, it should also be considered that the interaction of an alkali cation with the oxygen of a carbonyl group leads to a lowering of the energy of the lowest unoccupied orbital (frontier orbital) of the carbonyl group (Anh, 1976). This certainly contributes to facilitating the insertion reaction.

A related strong acceleration of a migratory insertion by the action of the Lewis acid $AlBr_3$ has been found by Shriver and his school (Butts et al., 1980). In the absence of excess CO an adduct was formed and structurally characterized (X-ray diffraction):

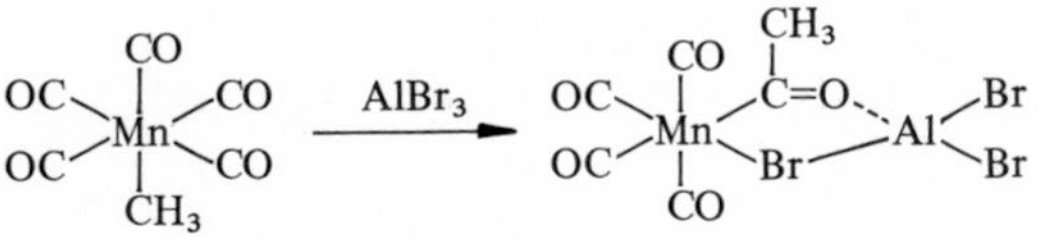

The Lewis acid not only interacts with the acetyl oxygen, but also forms a bridge bond to the metal, the bromine occupying the coordination site vacated by the migratory insertion reaction. The result is a five-membered ring providing the Mn center with a nearly perfect octahedral environment. The interaction Mn−Br is quite strong as judged from the bond length (257.9 pm) which is only slightly longer than that in $[BrMn(CO)_4]_2$ (252.6 pm). The C−O bond distance, 130 pm, in the aluminum coordinated acyl ligand is longer than that in simple η^1-acetyl ligands, which range from 119 to 121 pm. In the presence of CO the five membered ring opens up, and the pentacarbonyl complex

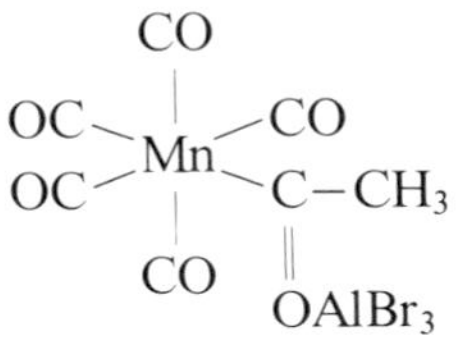

can be identified spectroscopically.

5.3.4 Absence of Acyl-to-CO Migration

One may ask why the octahedral acetyl complex formed in reaction (5.16) is not able to repeat the insertion reaction, leading to successive carbon-carbon bond formation. An attempt to prepare the model complex $CH_3CCMn(CO)_5$

by carbonylation of $CH_3CMn(CO)_5$ at 258 atm CO and 80 °C failed (Casey

et al., 1976). In fact such multiple CO insertion has never been observed in any CO hydrogenation or related process. Casey et al. (1976) were able to synthesize the model complex in a different way, reacting $NaMn(CO)_5$ with pyruvoyl chloride:

$$CH_3CCCl + NaMn(CO)_5 \rightarrow CH_3CCMn(CO)_5.$$

Thermal decomposition of the pyruvoyl manganese complex leads to an equilibrium mixture of $CH_3CMn(CO)_5$ and $CH_3Mn(CO)_5$; however, the decomposition of CH₃CCMn(CO)₅ is 21 times slower (at 75 °C) than the

decomposition of $CH_3CCMn(CO)_5$ is 21 times slower (at 75 °C) than the

decarbonylation of CH₃CMn(CO)₅. Evidently, an inherent instability of the

pyruvoyl complex is not the cause of its failure to form by acetyl migration, although Berke and Hoffmann (1978) suggested a weak C—C bond between the two carbonyl groups on theoretical grounds. But their calculations pointed also to an increased activation energy for the acetyl migration as compared to methyl migration. This reason for the difficulty of acetyl migration is substantiated by an amazingly short, strong metal-carbon bond found in several acetyl complexes (X-ray diffraction). Thus, the square planar nickel complex trans-CH₃CNiClL₂ (L = P(CH₃)₃) has a very short Ni—C bond (184 pm) as

compared with a typical Ni—C σ bond (194 pm), and also the frequency of the Ni—C stretching of the acyl complex is displaced to higher values as compared to those in methyl complexes. This has been considered as a good evidence for substantial dπ—pπ interaction, in other words for a partial double bond character of the Ni—C bond in the acetyl complex (Klein and Karsch, 1976). The rhodium acyl complex mentioned in Sect. 5.3.2 (Eq. 5.19) has also a Rh—C distance which is shorter than usual Rh—C σ bonds (Cheng et al., 1977). In other cases the incapability of an acyl ligand to undergo further CO insertion may be caused by π-coordination (dihapto coordination) of the acyl ligand. The X-ray structures of dihapto acyl complexes have been reported for group IV, V, VI and VIII metals (Collman and Hegedus, 1980); one example is

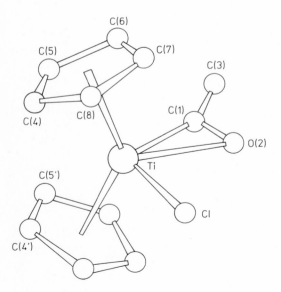

Fig. 5.1. Molecular structure of (η⁵-C₅H₅)₂Ti(C(O)CH₃)Cl (Fachinetti et al., 1977). Reproduced by permission of The Royal Society of Chemistry

shown in Fig. 5.1. Again a shortening of the Ti−C (acetyl) bond (207 pm) compared to the average Ti−C σ bond (214 pm) indicates strong bonding.

5.3.5 CO Insertion into Metal−H Bonds

The theoretical work of Berke and Hoffmann (1978) compares the migratory insertion of CO into metal−H bonds with that into metal−R bonds (R = alkyl). It is pointed out that the major energy change occurring in this type of reaction is associated with the conversion of the metal-alkyl, or metal-hydrogen, σ bond to the acyl π-type orbital (see Fig. 5.2). Consequently, the

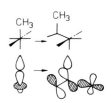

Fig. 5.2. Conversion of a metal-methyl to a metal-acyl group: representation of the relevant orbitals, according to Berke and Hoffmann (1978). Reproduced by permission of the American Chemical Society

higher the energy of the migrating group the easier the migratory process. In other words, the better σ-donor is the better migrating group. According to the spectrochemical series of ligands, the donor capacity of H^- and CH_3^- is approximately the same (see e.g. Henrici-Olivé and Olivé, 1977). In the calculations of Berke and Hoffmann, the hydrogen 1 s orbital is of lower energy than the methyl lone pair. However, there is a compensating stabilization in the orbital that develops into the new C−H bond in the case of hydride migration. As a result, the calculated activation energy is even somewhat smaller for formyl than for acetyl formation. Though the absolute values of activation energies resulting from such computations may be somewhat questionable, Berke and Hoffmann believe in a low barrier to the hydride migration.

The scarcity of direct observation data of formyl formation by the migratory route has been amply discussed in Chap. 4.2.2. Berke and Hoffmann suggest that formyl intermediates may in fact be formed in many cases, but may be susceptible to homolytic or heterolytic cleavage. Actually, the hydride transferring power of a formyl complex in the presence of a suitable acceptor has been reported (Gladysz and Selova, 1978).

Byers and Brown (1977) compared the rate of incorporation of ^{13}CO into $CH_3Mn(CO)_5$ (Eq. 5.16) with that of the exchange of a ^{13}CO ligand for a ^{12}CO in $HMn(CO)_5$, (Eq. 4.10). These reactions are assumed to take place via alkyl and hydride migration to a CO ligand, respectively, and comparable rate constants for the two migrations were deduced from kinetic measurements. However, whereas the acyl complex (Eq. 5.16) is relatively stable, the formyl ligand decarbonylates rapidly, giving rise to a ligand exchange (Eq. 4.10).

5.3.6 Migratory Insertions involving Carbene Ligands

Migratory insertions of various groups (H, alkyl, chloride) to a carbene ligand appear to be quite facile, even more so than the corresponding reactions involving carbon monoxide (Collman and Hegedus, 1980). It should be noted that migratory insertion according to

$$M=CH_2 \rightarrow M-CH_2-X \qquad\qquad (5.23)$$
$$\overset{|}{X}$$

is inferred by analogy with the carbonyl case. In fact insertion by intermolecular addition (generally within a binuclear intermediate) is also known in carbene complexes. However, in some cases the reaction is clearly intramolecular. Thus, Thorn and Tulip (1981) demonstrated the formation of an ethyl ligand through the migration of a methyl group towards a carbene group, cis-coordinated to the same metal center. The carbene ligand was created in situ by reacting a methyl(methoxymethyl)iridium complex with $BrCH_2OCH_3$:

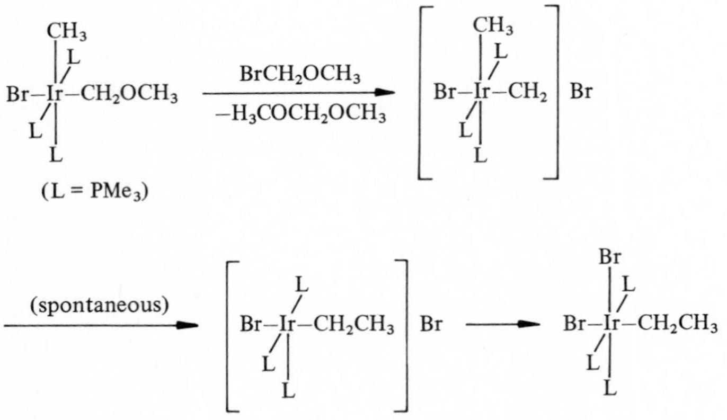

$$
\begin{array}{c}
CH_3 \\
\overset{|}{\underset{}{}} L \\
Br-Ir-CH_2OCH_3 \\
\overset{}{L} \overset{|}{\underset{L}{}}
\end{array}
\xrightarrow[-H_3COCH_2OCH_3]{BrCH_2OCH_3}
\left[
\begin{array}{c}
CH_3 \\
\overset{|}{\underset{}{}} L \\
Br-Ir-CH_2 \\
\overset{}{L} \overset{|}{\underset{L}{}}
\end{array}
\; Br
\right]
$$

$$(L = PMe_3) \qquad\qquad (5.24)$$

$$
\xrightarrow{(spontaneous)}
\left[
\begin{array}{c}
L \\
\overset{}{\diagup} \\
Br-Ir-CH_2CH_3 \\
\overset{}{L} \overset{|}{\underset{L}{}}
\end{array}
\; Br
\right]
\longrightarrow
\begin{array}{c}
Br \\
\overset{|}{\underset{}{}} L \\
Br-Ir-CH_2CH_3 \\
\overset{}{L} \overset{|}{\underset{L}{}}
\end{array}
$$

The structure of the final ethyliridium complex was confirmed by X-ray crystal structure determination. Similar insertion reactions into metal-methyl bonds have been reported by Maitlis and coworkers (Isobe et al., 1981) and by Cooper and coworkers (Hayes et al., 1981) with bridging (μ-methylene) and terminal (methylidene) carbene groups, respectively.

Threlkel and Bercaw (1981) observed the migratory insertion of a carbene ligand into a metal−H bond, in a binuclear Nb/Zr complex:

$$Cp_2(H)Nb=CHOZr(H)Cp_2^* + CO \rightarrow Cp_2(CO)NbCH_2OZr(H)Cp_2^* \qquad (5.25)$$

($Cp = \eta^5\text{-}C_5H_5$; $Cp^* = \eta^5 C_5(CH_3)_5$; Zr is bonded to O in both complexes). The rate of this reaction was compared to those of similar alkyl-to-carbene migrations, and it was found that the relative migratory aptitude decreases in the series $H \gg CH_3 > CH_2C_6H_5$.

On the other hand, also the insertion of CO into metal-μ-methylene bonds has been observed. Morrison et al. (1983) found the following reaction:

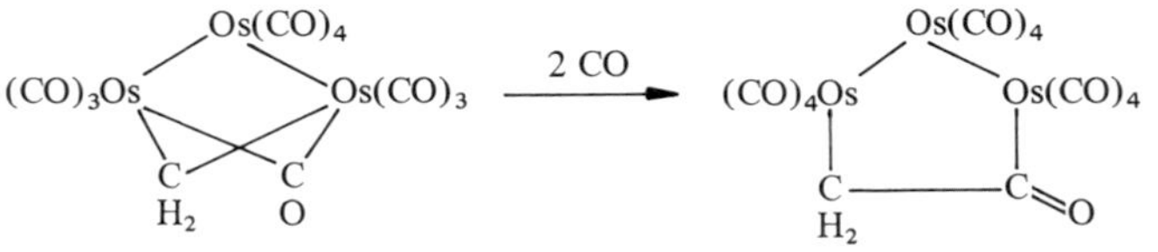

$$(5.26)$$

The ketene complex forms slowly if the μ-methylene complex is exposed to CO in solution. The trimetallacyclopentanone structure of the reaction product has been determined by X-ray diffraction.

The stretching frequency of the ketene CO is 1573 cm^{-1}. Labeling experiments have shown that the ketene carbonyl is one of the ligands initially present in the μ-methylene complex. The ketene ligand is very reactive; it gives with water acetic acid and with methanol methyl acetate.

A similar insertion of CO into a metal-μ-methylene bond has been reported by Lin et al. (1983):

$$Cp(CO)_2Ru-CH_2-Ru(CO)_2Cp \xrightarrow{CO} Cp(CO)_2Ru-\overset{\overset{\displaystyle O}{\|}}{C}-CH_2-Ru(CO)_2Cp.$$

5.3.7 Olefin Insertion

As mentioned at the beginning of this Chapter, insertion of olefins in $M-R$ bonds (R=H or alkyl) is only of secondary interest in our present context. These reactions are, however, of greatest importance for the initiation (R = H) and propagation (R = growing chain) of the oligomerization and polymerization of olefins with Ziegler type catalysts (see e.g. Henrici-Olivé and Olivé, 1974), and for the hydrogenation of unsaturated compounds (see e.g. James, 1973).

The insertion of an olefin is assumed to be a concerted reaction, with a more or less polar cyclic transition state involving simultaneous bond breaking and bond making:

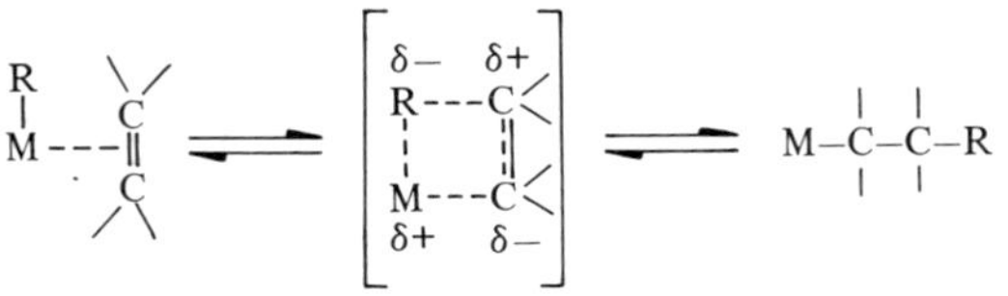

$$(5.27)$$

The reason for the assumption of a concerted reaction is as discussed for the CO insertion in Sect. 5.3.1: low activation energy compared to the dissociation energy of the bonds to be broken in the process; negative activation entropy. As an example, the activation parameters for the insertion of ethylene into the metal-carbon bond of an ethylrhodium (III) complex are (Cramer, 1965): Arrhenius overall activation energy $E_a = 72$ kJ/mol, $\Delta H^* = 69.5$ kJ/mol, $\Delta F^* = 95$ kJ/mol, $\Delta S^* = -81.4$ kJ/mol. This compares to the strengths of the involved bonds as follows: $M-R$ bonds are generally in the range $160-350$ kJ/mol (Connor, 1977); the strength of the coordinative rhodium-ethylene bond has been estimated as ≤ 130 kJ/mol (Cramer, 1972; although this is an upper limit, it indicates that metal-olefin bonds are not weak); finally, the opening of a carbon-carbon double bond requires 264 kJ/mol.

Green (1978) suggested that the insertion of an olefin into a $M-R$ bond ($R = $ alkyl) might proceed by a different mechanism, involving α-elimination (cf. Sect. 5.4.2) of a hydrogen, leading to a metal-carbene bond; the carbene and the approaching olefin would form a metallacycle which decomposes by reductive elimination of H and one end of the metallacycle:

$$M-CH_2-R \rightleftharpoons \underset{\overset{|}{M}=CH-R}{\overset{H}{}} \quad \overset{>C=C<}{\longrightarrow}$$

(5.28)

$$\underset{\overset{\overset{H}{|}}{M}-C^1H-R}{\overset{H}{|}} \longrightarrow M-\overset{|}{C^3}-\overset{|}{C^2}-C^1H_2-R$$

However, Evitt and Bergman (1979) have demonstrated by labeling experiments that in the case of the cobalt complex $(\eta^5\text{-}C_5H_5)Co(CH_3)_2(PPh_3)$ incorporation of ethylene takes place according to Eq. (5.27) (insertion) and not Eq. (5.28) (α-elimination/metallacycle). The perdeuterated Co complex reacts with C_2H_4 stoichiometrically to $CD_3CH=CH_2$ and CD_3H. These are the products to be expected from insertion, followed by β-H elimination (cf. Sect. 5.4.1) and reductive elimination, according to:

$$CpCo(PPh_3)(CD_3)_2 \underset{\overset{-PPh_3}{}}{\rightleftharpoons} CpCo(CD_3)_2 \underset{\overset{C_2H_4}{}}{\rightleftharpoons} CpCo \cdots \| \overset{CD_3}{\underset{CD_3}{}} \overset{CH_2}{\underset{CH_2}{}} \longrightarrow$$

$$CpCo \overset{CH_2-CH_2-CD_3}{\underset{CD_3}{}} \longrightarrow CpCo \cdots \| \overset{H \ CD_3}{\underset{CD_3}{\overset{C}{}}} \overset{C_2H_4, PPh_3}{\longrightarrow}$$

$$CpCo(PPh_3)(C_2H_4) + CD_3H + CH_2=CH-CD_3$$

These results are not compatible with the α-elimination mechanism, where CD_4 and a d^2 propene molecule should have been the outcome.

While this is a clear-cut demonstration of olefin incorporation by insertion, it was not considered definite proof for catalytic olefin incorporation (as in Ziegler type polymerization) proceeding according to the same insertion mechanism, until the more recent work of Grubbs and coworkers (Soto et al., 1982). One important difference between the α-elimination mechanism and that depicted in Eq. (5.27) is the involvement of hydrogen migration in the former. This would imply a large primary kinetic isotope effect in the range of $k_H/k_D \simeq 3$, in the case of mechanism (5.28). However, Grubbs found k_H/k_D very close to unity when copolymerizing C_2H_4 and C_2D_4 with a Ziegler type catalyst. These data strongly support the insertion mechanism Eq. (5.27), which does not involve hydrogen migration in the growth step of the polymerization.

Carbene ligands, on the other hand, are more probable to insert olefins via a metallacycle intermediate: Thorn (1982), in a reaction similar to that described in Eq. (5.24), used a methoxymethyl ligand on an Ir complex as a convenient precursor for the in situ preparation of a carbene ligand, in the presence of a coordinated ethylene molecule. The outcome of the reaction is a cationic hydridoallyl complex. The most plausible mechanism involves a metallacycle:

$$
\begin{array}{c}
L_3Ir-CH_2OCH_3 \\
\overset{..}{\underset{>C=C<}{}}
\end{array}
\xrightarrow[-EOCH_3]{+E^+}
\left[
\begin{array}{c}
L_3Ir=CH_2 \\
\overset{..}{\underset{>C=C<}{}}
\end{array}
\right]^+
\longrightarrow
\left[
\begin{array}{c}
H \\
| \\
L_3Ir-CH \\
| \quad | \\
-C-C- \\
| \quad |
\end{array}
\right]^+
$$

$$
\longrightarrow
\left[
\begin{array}{c}
H \\
| \\
L_3Ir- \triangleright
\end{array}
\right]^+
$$

(L = PMe$_3$, E$^+$ = trimethylsilyl trifluoromethanesulfonate).

In other cases, substituted olefins react with carbene complexes forming cyclopropanes (see e.g. Casey, 1979):

$$
(CO)_4W\!=\!C\!\begin{array}{c}-C_6H_5 \\ \diagdown C_6H_5\end{array}\!\!\underset{R}{\diagup}
\;\rightleftharpoons\;
(CO)_4W\!-\!\begin{array}{c}C_6H_5\\ \diagup \quad -C_6H_5\end{array}\!\underset{R}{\;}
\;\longrightarrow\;
\begin{array}{c}H \quad \quad C_6H_5\\ \diagdown \!\triangle\! \diagup \\ R \quad \quad C_6H_5\end{array}
\tag{5.29}
$$

A high yield reaction of this last type, involving an iron carbene complex and an olefin has been reported by Brookhart et al. (1981):

$$
(\eta^5-C_5H_5)(CO)_2Fe^+\!=\!C\!\begin{array}{c}-H \\ \diagdown CH_3\end{array}
\;\xrightarrow{\;>C=C<\;}\;
\triangleright\!\!-CH_3
$$

5.3.8 Coinsertion of Ethylene and CO

Kaesz and his school (Kampe et al., 1983) have reported the following interesting reaction of a hydrogen bridged Ru cluster with C_2H_4 and CO:

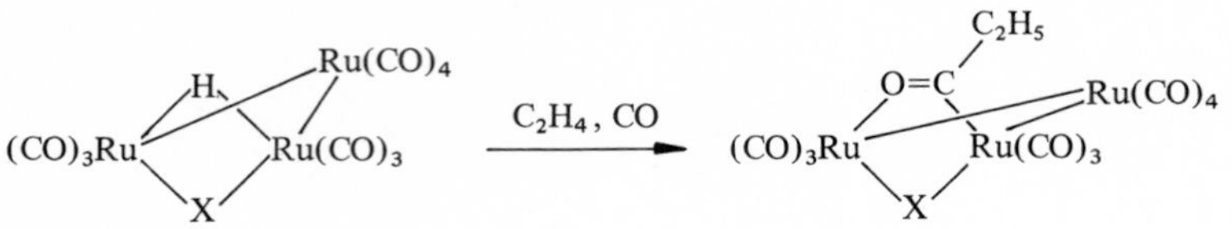

$$(5.30)$$

$(X = Cl, Br, I; 26\,^{\circ}C$ in hexane; $p_{C_2H_4} = 3$ atm, $p_{CO} = 1$ atm).

Apparently, the hydride has inserted an ethylene molecule, followed by insertion of a CO molecule. The µ-acetyl complex is characterized by NMR and IR spectroscopy and elemental analysis.

Even a step further is the alternating copolymerization of C_2H_4 and CO on a mononuclear palladium complex, found by Sen and Lai (1982). The compounds $[Pd(CH_3CN)_4(PPh_3)_n][BF_4]_2$ (n = 1−3) catalyze this copolymerization at $25\,^{\circ}C$ in $CHCl_3$, if exposed to a mixture of CO ($\simeq 24$ atm) and C_2H_4 ($\simeq 24$ atm). The reaction is relatively slow (a combined turnover of CO and C_2H_4 of 300 after one day). The 1:1 copolymer has been characterized by elemental analysis and ^{13}C NMR. It is a white solid with a melting point of $260\,^{\circ}C$ and is virtually insoluble in all common organic solvents, presumably due to high crystallinity. In the absence of CO, the catalyst dimerizes ethylene. This points to a hydride as the active species (as in Eq. 5.30); however, since in this case no bridging is involved, the acyl group formed after the successive insertion of ethylene and CO is not stabilized, but is able to repeat the alternating insertion:

$$L_nPd-H \xrightarrow{C_2H_4} L_nPd-C_2H_5 \xrightarrow{CO} L_nPd-\underset{\underset{O}{\|}}{C}-C_2H_5 \xrightarrow{C_2H_4} \cdots$$

The origin of the hydride is not clear. The insertion of CO into the M−R bond (R = alkyl) is evidently considerably faster than the insertion of a second C_2H_4 (the general validity of this statement permits the hydroformylation of olefins, cf. Chap. 10). After the insertion of a CO, on the other hand, no second CO can be inserted (cf. Sect. 5.3.4). Obviously, the slow step is the insertion of C_2H_4 into the acetyl-metal bond. (In the hydroformylation case this step does not take place because of the presence of a very active hydride that reduces the acetyl ligand to aldehyde, see Chap. 10.)

5.3.9 Insertion of Formaldehyde

Anticipating some material that will be treated in much more detail in later Chapters, we want to discuss briefly the insertion of formaldehyde into metal−H bonds. Formaldehyde coordinatively bonded to a metal center is considered as a key intermediate, and its insertion into a metal−H bond as a key step, in the mechanisms of the catalytic heterogeneous production of hydrocarbons, (Chaps. 7, 9), as well as the homogeneous formation of glycol, higher polyols, formates, etc., (Chap. 10).

In contrast to CO which is generally bonded "end-on" to a metal, formaldehyde has been reported to be π-bonded (X-ray diffraction; see Table 4.1). Thus, while the insertion of CO in a M−H bond (or M−alkyl bond) occurs unequivocally by reaction of the hydrogen (or the alkyl) ligand with the carbon side of the molecule

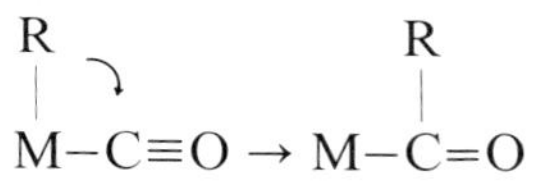

$$\begin{array}{ccc} R & & R \\ | \;\diagdown & & | \\ M-C{\equiv}O & \rightarrow & M-C{=}O \end{array}$$

(R = H or alkyl), the insertion of formaldehyde can, in principle, take place in either of the two modes shown in Eqs. (5.31) and (5.32):

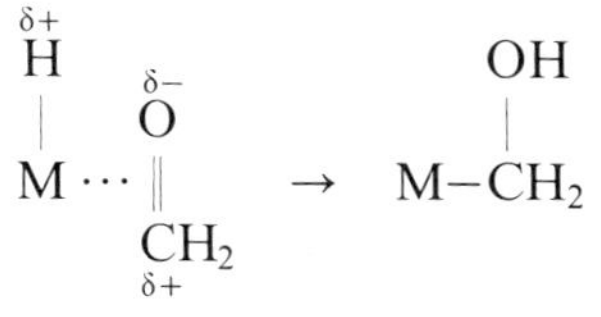

$$\begin{array}{cccc} \overset{\delta+}{H} & & & \\ | & \overset{\delta-}{O} & & OH \\ M \cdots \| & & \rightarrow & | \\ & CH_2 & & M-CH_2 \\ & \underset{\delta+}{} & & \end{array} \qquad (5.31)$$

$$\begin{array}{cccc} \overset{\delta-}{H} & & & \\ | & \overset{\delta+}{CH_2} & \rightarrow & M-OCH_3 \;. \\ M \cdots \| & & & \\ & \underset{\delta-}{O} & & \end{array} \qquad (5.32)$$

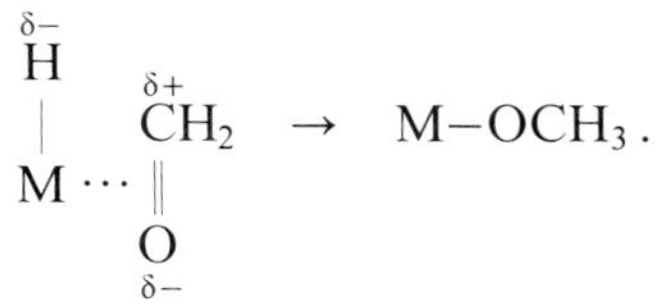

If at this stage hydrogenolysis takes place (by reductive elimination with a metal hydride), methanol will be the only product, and no distinction between the two insertion modes is possible. Further reaction steps must follow in order to make a decision.

Roth and Orchin (1979) carried out the reaction of formaldehyde and HCo(CO)$_4$, under 1 atm of CO, at 0 °C in CH$_2$Cl$_2$, and found up to 90% (referred to HCo(CO)$_4$ consumption) of glycol aldehyde. The plausible reaction mechanism is starting with Eq. (5.31), followed by insertion of CO, and intermolecular reductive elimination with $HCo(CO)_4$:

$$(CO)_4CoH + H_2CO \rightarrow (CO)_4Co\overset{OH}{\underset{|}{C}}H_2 \rightarrow (CO)_3Co\overset{O\ \ OH}{\underset{\|\ \ |}{C}}-CH_2$$

$$\xrightarrow[CO]{(CO)_4CoH} H\overset{O\ \ OH}{\underset{\|\ \ |}{C}}-CH_2 + Co_2(CO)_8.$$

Since formaldehyde has a strongly polar group, with the negative charge on the oxygen, it follows that the hydride has functioned as a weak acid in the reaction, as it has been observed for many metal carbonyl hydrides in polar solvents (see Table 2.2).

Vaughn and Gladysz (1981) succeeded in isolating the α-hydroxyalkyl complex resulting from insertion of a phosphine substituted benzaldehyde into the metal-hydrogen bond of $(CO)_5MnH$. In this case, the resulting complex was stabilized by an internal ring formation involving the metal and the phosphine substituent:

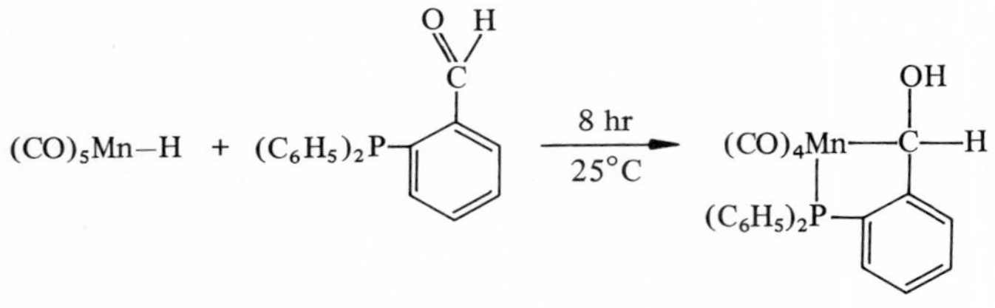

The insertion of formaldehyde (and higher aldehydes) into the metal−H bond of a rhodium octaethylporphyrin hydride was also observed (Wayland et al., 1982). The generated α-hydroxyalkyl ligand was characterized by ^{1}H NMR.

$$Rh-H + H-\overset{H}{\underset{\|O}{C}}-R \rightarrow Rh-\overset{H}{\underset{OH}{C}}-R.$$

The compounds are stable and could be isolated. As in the case of CO insertion with the same Rh−H complex (cf. Sect. 4.2.2), migration of the hydride ligand from one axial position to the other may be involved.

In principle, it is conceivable that the other insertion mode (Eq. 5.32) takes place under different conditions (other metal, ligand environment, valency state, solvent, pressure, temperature) and indeed there is (indirect) evidence for it. Bradley (1979) reported that the reaction of $CO + H_2$ with a ruthenium catalyst (1300 atm, 268 °C, in solution) led to methanol and methyl formate. Anticipating here again the intermediate formation of coordinatively bonded formaldehyde (cf. Chaps. 7, 9), the production of formate indicates reaction 5.32, followed by CO insertion and reductive elimination with a hydride:

$$
\text{Ru}-\text{OCH}_3 \xrightarrow{\text{CO}} \text{Ru}-\overset{\overset{\text{O}}{\|}}{\text{C}}-\text{OCH}_3 \rightarrow \text{HCOCH}_3 .
$$

Thus, in homogeneous medium, both addition modes appear possible, although that according to Eq. (5.31) seems to be the more frequent one.

In heterogeneous systems, no direct evidence is available with regard to the two insertion modes (5.31) and (5.32). However, indirect evidence (i.e. drawing conclusions from the reaction products, cf. Chaps. 8 and 9) shows that both modes are possible, depending on catalyst system and conditions.

5.4 Hydrogen Elimination Reactions

5.4.1 β-Hydrogen Elimination

Transition metal complexes with σ bonded organic ligands having hydrogen attached to the β-carbon tend to undergo C−H bond rupture forming a metal hydride. The organic ligand leaves the complex with an olefinic end group.

$$
\text{M}-\text{CH}_2\text{CH}_2\text{R} \rightleftharpoons \begin{bmatrix} \text{H}----\text{CHR} \\ | \quad\quad \| \\ \text{M}----\text{CH}_2 \end{bmatrix} \rightleftharpoons \text{M}-\text{H} + \text{CH}_2{=}\text{CHR} \tag{5.33}
$$

In Ziegler type polymerization of olefins, for instance, the β-hydrogen elimination (or β-H transfer) determines the molecular weight of the polymer; if β-H transfer takes place after only two growth steps, dimerization of the olefin occurs. The tendency towards β-hydrogen elimination depends upon the metal (group VIII metals > metals from the left hand side of the Periodic Table), on the valency state (e.g. Ti(IV) > Ti(III)) and on the ligand environment (see Sect. 5.5.1).

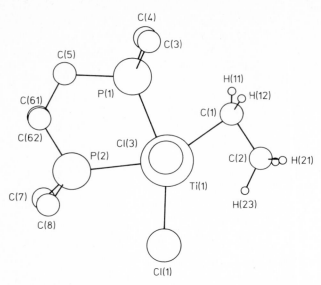

Fig. 5.3. Crystal structure of $(C_2H_5)Ti(Me_2PCH_2CH_2PMe_2)Cl_3$ showing the distortion of the ethyl group and the proximity of one β-H and metal. The Cl(2) atom (not shown) is located symmetrically on the Cl(3)-Ti axis below the plane (Dawoodi et al., 1982). Reproduced by permission of The Royal Society of Chemistry

The reaction cannot take place unless an unoccupied coordination site in cis-position to the alkyl ligand is available, indicating that the hydrogen has to approach the metal in a cyclic transition state such as formulated in Eq. (5.33) above. Formally, β-elimination is the reverse of olefin insertion (Eq. 5.27, with R=H), and actually the reaction is often reversible.

A kinetic isotope effect has been observed for the β-hydrogen elimination in the square planar iridium complex $C_6H_{13}CH(D)CH_2Ir(CO)(PPh_3)_2$ (Evans et al., 1974). From the ratio of deuterated to non-deuterated octene, a value of $k_H/k_D = 2.28$ was determined. This implies a primary kinetic isotope effect, involving hydrogen (or deuterium) in the β-position, referred to the metal center. The relatively low value of k_H/k_D* accords with a cyclic transition state involving simultaneous bond breaking and bond making (Egger, 1969).

An illustrative example of β-H abstraction where the intermediate hydrido-olefin complex as well as the hydride (after the expulsion of the olefin) could be isolated and characterized, was reported by Alt and Eichner (1982):

$$Cp(CO)_3WCH_2CH_3 \underset{+CO}{\overset{h\nu,\,-CO}{\rightleftarrows}} [Cp(CO)_2WCH_2CH_3] \rightarrow Cp(CO)_2W\cdots \overset{CH_2}{\underset{\underset{H}{|}}{\|}}_{CH_2}$$

$$\xrightarrow{h\nu,\,+CO} Cp(CO)_3WH + C_2H_4.$$

---

* The theoretical value for a linear transition state is $k_H/k_D \simeq 7$ at 25 °C. There exist reasons other than cyclic transition states for lower values (Westheimer, 1961).

The rate determining step is the reversible CO cleavage from the starting ethyl complex, necessary to provide the required coordination site.

As mentioned above, Ti(IV) has a particular propensity for β-H abstraction, and Ti(IV)-alkyl complexes are usually unstable. However, Green and his school (Dawoodi et al., 1982a) succeeded in stablilizing a Ti(IV)-ethyl complex with the bidentate phosphine ligand $Me_2PCH_2CH_2PMe_2$ (dmpe). The crystal structure of the complex $Ti(C_2H_5)Cl_3$ (dmpe) is shown in Fig. 5.3. It shows that one of the β-hydrogens of the ethyl ligand is very close to the metal in an almost bridge bonding position, although without occupying a separate coordination site in the distorted octahedral structure. This interaction causes a modification of the Ti−C−C angle from the usual sp^3 tetrahedral angle of 109° to 86°. This kind of interaction can be assumed to be particularly strong in d^0 complexes with electron withdrawing ligands, as in the present case (three chlorine ligands); however, to a lesser degree it is most probably present in other transition metal-alkyl species, and may be viewed as an incipient hydrogen elimination.

5.4.2 α-Elimination of Hydrogen

Since Michman and Zeiss (1968) first hinted to the possible involvement of α-hydrogen elimination with simultaneous formation of a carbene ligand in the decomposition of transition metal alkyls, growing evidence for the importance of this type of reaction in a number of organometallic reactions has evolved.

The prototype of this reaction may be formulated as

$$L_nM-CH_3 \rightleftarrows L_nM{\overset{H}{=}}CH_2. \tag{5.34}$$

Pu and Yamamoto (1974) found polydeuterated methanes when reacting L_3MCH_3 (M = Co, Rh; L = PPh$_3$) with D_2 and explained this finding by assuming rapid reversible α-hydrogen abstraction, followed by exchange with deuterium from the gas phase, and back-migration of the deuterium to the carbene ligand to restore the methyl group (CH_2D) which then leaves the complex by reductive elimination with a deuteride ligand, e.g.

$$Co-CH_3 \rightleftarrows Co{\overset{H}{=}}CH_2 \xrightarrow{D_2} Co{\overset{D}{=}}CH_2 \rightleftarrows Co-CH_2D$$

$$\xrightarrow{D_2} Co-D + CH_2D_2 \tag{5.35}$$

Canestrari and Green (1982, and earlier references therein) summarized extensive NMR and deuterium labeling evidence for α-H elimination in the case of a methyltungsten complex. Green has found that $[(\eta^5\text{-}C_5H_5)W(CH_3)]^+$ forms two different adducts with phosphines:

$$\underset{}{Cp_2\overset{+}{W}}\underset{CH_3}{\overset{PR_3}{<}} \quad \text{and} \quad Cp_2W\underset{H}{\overset{\overset{+}{C}H_2PR_3}{<}}.$$

The collected evidence indicates a facile equilibrium between two forms of the starting complex, produced by reversible α-H elimination:

$$Cp_2\overset{+}{W}\text{-}Me \rightleftarrows Cp_2\overset{+}{W}\underset{H}{\overset{CH_2}{<}} \qquad (5.36)$$

each of which gives rise to one of the adducts.

A theoretical study of equilibrium (5.36) by Hoffmann and coworkers (Goddard et al., 1980) indicates that little activation energy should be required, and that both species have low lying empty orbitals available for further reaction.

To strengthen his case of α-H elimination, Green has determined the X-ray crystal structure of the methyl equivalent of the complex represented in Fig. 5.3 (Dawoodi et al., 1982b). One of the methyl hydrogens is reported to be tilted towards the metal center with a M−H distance of 203 pm (well within the sum of the van der Waals radii) and a M−C−H angle of 70°. Again a situation of incipient α-elimination appears to be indicated. Although in general neutron diffraction is considered a necessity for the accurate location of the crucial hydrogen (Goddard et al., 1980), the authors affirm that, due to the low thermal motion of the methyl group, the hydrogens are clearly resolved in this particular X-ray pattern.

Certain alkylidene ligands have been found by Schrock and his school to have also a distorted structure, with the α-hydrogen almost in a bridging position between the metal and the α-carbon (Churchill et al., 1982, and references therein):

$$L_nM=C\underset{H_\alpha}{\overset{R}{<}} \qquad (M = Ta, Nb, W).$$

In one case, complete transfer of the α-H to the metal has been observed:

$$(dmpe)_2(I)Ta=CHCMe_3 \rightleftarrows (dmpe)_2(I)Ta\equiv CCMe_3$$
$$| \\ H$$

(dmpe = $(CH_3)_2PCH_2CH_2P(CH_3)_2$). At 60°, the alkylidene and the alkylidyne hydride complexes are present in a 1:1 mixture (in toluene solution), and they interconvert rapidly on the NMR time scale.

Elimination of α-H from a bridging methyl group in an osmium cluster was observed by Calvert and Shapley (1977), who found the cluster (μ-CH_3) (μ-H) $Os_3(CO)_{10}$ in equilibrium with a μ-methylidene cluster:

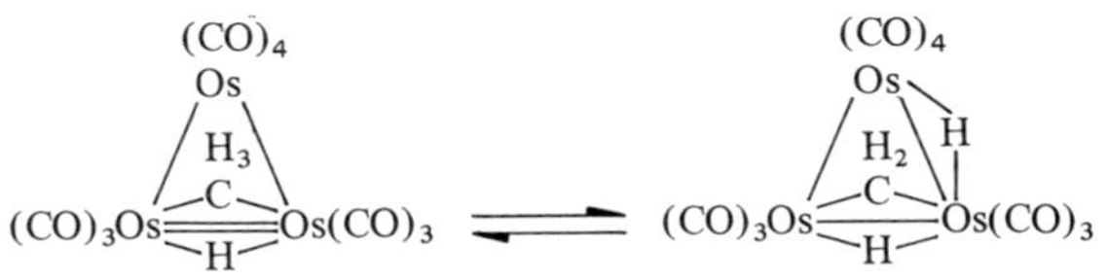

(5.37)

A detailed NMR study of partially deuterated forms of the complexes indicated that the bridging methyl ligand is not symmetric (as it is generally accepted for, e.g. $Al_2(CH_3)_6$), but unsymmetric with one hydrogen in close interaction with one of the metals (Calvert and Shapley, 1978):

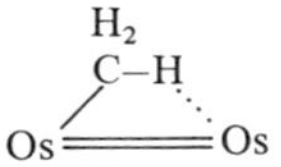

Thus, α-hydrogen élimination appears by now as a fairly well established reaction pattern.

5.4.3 γ and δ Elimination of Hydrogen

The formation of a metallacycle by γ-H abstraction from the neopentyl ligand has been reported by Tulip and Thorn (1981) for an iridium complex:

$$[Ir(PMe_3)_4]Cl \xrightarrow[-PMe_3]{+LiCH_2CMe_3} [(PMe_3)_3IrCH_2CMe_3]$$

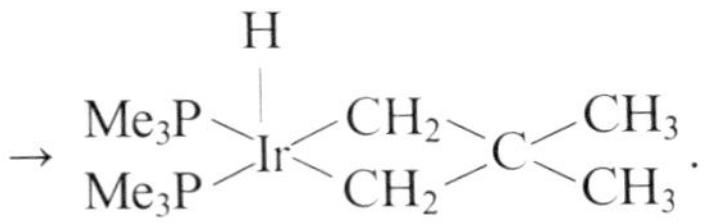

The assumed intermediate alkyl complex was not detected, it was, however, found in the Si analog prepared using $LiCH_2SiMe_3$ instead of $LiCH_2CMe_3$. Similar hydrogen abstraction from aromatic ligands, from γ- and δ-positions relative to the metal, have also been found, when $[Ir(PMe_3)_4]Cl$ was reacted with benzylmagnesium chloride in THF, or with $LiCH_2CMe_2Ph$ in hexane, respectively:

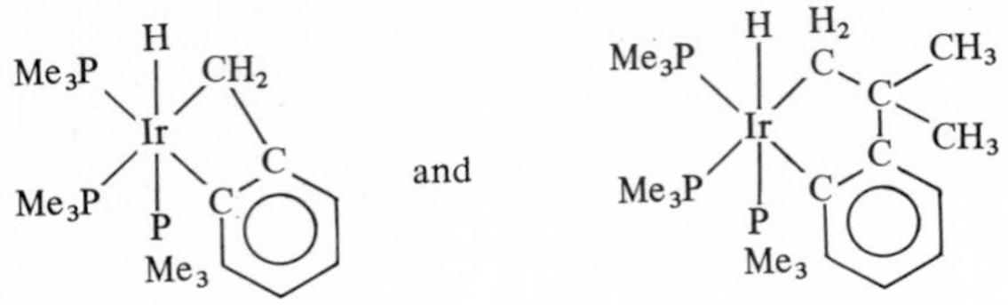

This type of synthesis derives from the propensity of Ir(I) centers to undergo facile oxidative addition reaction (cf. Sect. 5.1), and may be limited to relatively electron rich, later transition metals (for instance also Pt(II), Foley et al., 1980).

5.5 Ligand Influences

5.5.1 In Molecular Complexes

In the preceding Sections of this Chapter we have summarily represented by L_n those ligands which are not directly involved in a particular reaction. However, these ligands have frequently considerable indirect influence. Depending on their electron donating or accepting capacities, these ligands may vary the electronic structure in the entire complex by increasing the electron density in certain regions, decreasing it in others, thus changing the strength of those bonds actually involved in the reaction.

An illustrative theoretical treatment of ligand influences on bond strengths has been given by Zumdahl and Drago (1968). Using extended Hückel MO theory, the overlap population in the various bonds of the square planar platinum (II) complex trans-$PtL(NH_3)Cl_2$ was calculated, for varying L. The overlap population is a measure of the electron density in the bonding region of two atomic orbitals; a summation over all involved orbitals can be used as

Table 5.1. Overlap population for trans-$PtL(NH_3)Cl_2$ complexes (Zumdahl and Drago, 1968).

L	Pt−L	Pt−N	Pt−Cl
H_2O	0.241	0.377	0.391
NH_3	0.322	0.322	0.383
H_2S	0.429	0.324	0.374
PH_3	0.568	0.309	0.357
H	0.607	0.307	0.364

an indication of relative bond strength. Table 5.1 shows that the more strongly bonding the ligand L (i.e. the higher the overlap population in the Pt$-$L bond), the lower is the electron density in all other bonds. The effect is somewhat more pronounced in the trans-position, but clearly noticeable also in the cis-position.

The ligands NH_3 and Cl do not have empty π orbitals available for electron back donation. Hence the electronic influence of L on the Pt$-$N and Pt$-$Cl bonds may be assumed to operate via the σ orbital system. Qualitatively, the difference of the ligand effect on trans and cis position (which in practical experiments is often quite large) may be rationalized considering that all σ bonding orbitals ($d_{x^2-y^2}$, s, p_x or p_y) are shared by L and the trans ligand NH_3, but only two ($d_{x^2-y^2}$ and s) are shared by L and the cis ligands Cl.

For ligands having orbitals of the appropriate symmetry available (antibonding π^* in CO, d orbitals in phosphines, etc.), the influence of electrons in π orbitals is superimposed on the action via the σ orbital system. A π acceptor ligand withdraws electron density from π-bonds with other ligands, whereby the $d\pi$ orbitals of the metal act as a "conductor" for the electrons.

Such ligand effects are, in principle, accessible by vibrational spectroscopy (IR), crystallographic bond lengths determination, and other measurements on stable complexes, as well as by kinetic (rates) and thermodynamic (equilibria) measurements of model processes such as oxidative addition, reductive elimination, etc., in variable ligand environment.

The *oxidative addition* of hydrogen, for instance, involves electron donation from a metal dσ orbital to the antibonding σ^* orbital of H_2. After opening of the H$-$H bond, a pair of covalent, but polarized M$-$H bonds are formed (cf. Sect. 2.4). Evidently, high electron density at the metal center should be favorable, and in fact the donor character of the other ligands determines the ease of such reactions. For instance the equilibrium constant for the addition

$$RhClL_3 + H_2 \overset{K}{\rightleftarrows} Rh(H)(H)ClL_3 \qquad (5.38)$$

is the larger, the better the donor ability of the ligand L (L = P(p-$CH_3C_6H_4$)$_3$: K = 40 atm^{-1}; L = P(C_6H_5): K = 18 atm^{-1}; Tolman et al., 1974). On the other hand, acceptor ligands reduce the ease of oxidative additions. Thus, the relative rate of the reaction

$$IrX(CO)(PPh_3)_2 + H_2 \rightarrow H_2IrX(CO)(PPh_3)_2$$

decreases in the series for X: I ($> 10^2$) > Br (15) > Cl(1), the halogen ligand increasingly counteracting the donor ability of the two phosphine ligands in this series (Halpern and Chock, 1966).

In a *reductive elimination*, two metal-ligand σ bonds have to be broken, and donor ligands would be expected to be necessary also for this reaction in order to weaken these bonds. On the other hand, electron density donated from the metal to the ligands to be eliminated, has to be taken back by the metal, and this should be facilitated by the presence of electron withdrawing ligands. Phosphine ligands are good σ donors and moderately strong π acceptors (see Henrici-Olivé and Olivé, 1977). Abis et al. (1978) have shown that the complexes cis-PtH(CH₃)(Pr₃)₂ are stable only at very low temperature. At 25 °C, they decompose according to

$$\text{cis-PtH(CH}_3\text{)(PR}_3\text{)}_2 \rightarrow \text{Pt(PR}_3\text{)}_2 + \text{CH}_4 \tag{5.39}$$

with half-life times from 12 min to 4 h, whereby the rate of the reductive elimination increases with increasing π acceptor capability of the phosphine ligand (R = p-CH₃OC₆H₄ < p-CH₃C₆H₄ < p-C₆H₅ < p-ClC₆H₄). Apparently, the metal can channel the electron density taken up in the σ orbital system from the leaving ligands R and H into the π-accepting orbitals of the phosphines, via its d orbitals.

Migratory insertion of CO or alkenes into metal-alkyl or metal-hydrogen bonds involves breaking of these bonds. According to the data in Table 5.1, this should be facilitated by donor ligands, in particular if located in trans position to the migrating ligand. An illustrative example has been given by Cross and coworkers (Anderson and Cross, 1980; Cross and Gemmill, 1981). The CO insertion reaction

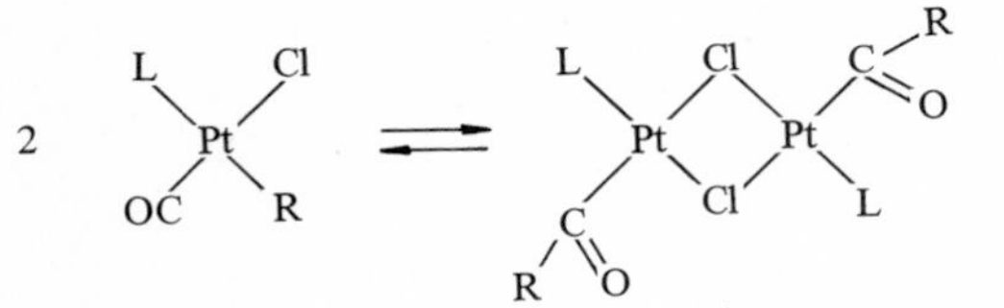

$$\tag{5.40}$$

(R = alkyl or aryl; L = tertiary phosphine or arsine) takes place easily at room temperature and is favored by donor ligands PEt₃ > PMe₂Ph > PMePH₂ > PPh₃). The two other isomers

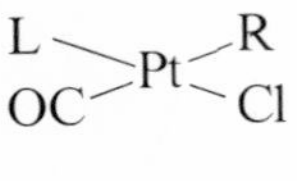

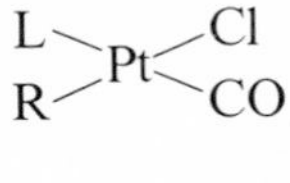

on the other hand, resist attempts to promote CO insertion (acceptor ligands CO and Cl, respectively, in trans position to R), although they equilibrate with each other in solution and are slowly converted into trans-PtRL(CO)Cl and the bridged dimer.

The favorable influence of donor ligands on the migratory insertion is also documented in ethylene polymerization, with the catalyst system

$$Cp_2TiEtCl + R^1R^2AlCl \longrightarrow \quad \begin{matrix} R^1 \\ \diagdown \\ Al \\ \diagup \\ R^2 \end{matrix} \begin{matrix} Cp \\ Cl \quad | \quad Et \\ \diagup \diagup \diagup \diagup Ti \\ Cl \diagdown \\ | \\ Cp \end{matrix} \qquad (5.41)$$

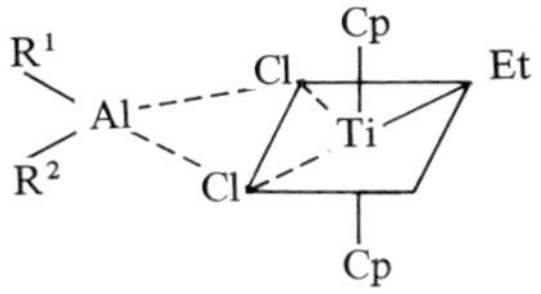

where the polymerization rate (rate of insertion of ethylene into the titanium-alkyl bond, after π coordination of the ethylene molecule to the free coordination site) increases with increasing donor capability of R^1 and R^2 ($R^1 = R^2 = Cl < R^1 = C_2H_5$, $R^2 = Cl < R^1 = R^2 = C_2H_5$) (Henrici-Olivé and Olivé, 1969). The improvement has been traced back in fact to a weakening of the metal-alkyl bond by determining the rate of breaking of this bond, in the absence of ethylene monomer, by magnetic measurements. The breaking of the Ti−ethyl bond in the complex represented on the right hand side of Eq. (5.41) is accompanied by reduction of Ti(IV) (diamagnetic) to Ti(III) (paramagnetic), and can easily be followed by measuring the magnetic susceptibility of the reaction solution. The rate increases with the donor ability of R^1 and R^2, parallel to the observed increase of polymerization rate. Note that in this case no π acceptor ligand is involved (as it was in Eq. 5.39), so that the influence of donor ligands on metal-carbon bond weakening becomes clearly evident. Interestingly, the rate of the reduction Ti(IV) $\rightarrow$ Ti(III) is second order with regard to the concentration of the starting Ti(IV) complex, and leads to the formation of C_2H_4 and C_2H_6. It was suggested that the reaction may represent an intermolecular reductive elimination, in the course of which a β-H of one leaving group passes to the other leaving group

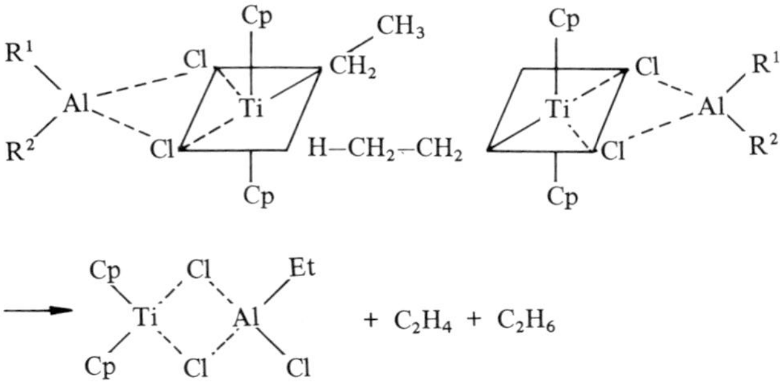

$$\longrightarrow \quad \begin{matrix} Cp \quad Cl \quad Et \\ \diagdown \diagdown \diagup \diagup \\ Ti \quad Al \\ \diagup \diagdown \diagdown \\ Cp \quad Cl \quad Cl \end{matrix} \quad + C_2H_4 + C_2H_6$$

Ligand influences on β-*hydrogen elimination* have been observed during the ethylene polymerization with Ziegler type catalysts. The elimination of a β-H is a molecular weight determining step; from a kinetic point of view, however, it is a chain transfer reaction, because the metal hydride formed in the reaction inserts monomer and thus continues the kinetic chain:

$$L_nM-CH_2CH_2-P \rightarrow L_nM-H + CH_2=CH-P$$
$$\searrow_{CH_2=CH_2}$$
$$\rightarrow L_nM-CH_2CH_3 . \qquad (5.42)$$

Table 5.2. β-H elimination in ethylene polymerization; $T = 20\,°C$, ethylene pressure 1 atm; $[Cr] = 2 \times 10^{-4}\,mol/l$ (Henrici-Olivé and Olivé, 1971).

Catalyst system	Polymerization rate[a]	β-H elimination rate[b]
Cr(acac)$_3$/Et$_2$AlCl	100	0
Cr(acac)$_3$/EtAlCl$_2$	15	2.3×10^{-2}

[a] monomer units grown per active site per minute;
[b] chain transfer steps per active site per minute.

In the absence of chain termination steps ("living polymers"), the frequency of chain transfer (β-H elimination) can be estimated from the degree of polymerization and the rate of polymerization with the equation

$$P_n = r_p/r_{tr} \tag{5.43}$$

(P_n = number average degree of polymerization, r_p and r_{tr} = rates of polymerization and of chain transfer, respectively). Table 5.2 shows data of the ethylene polymerization with Cr/Al catalyst systems. As expected, the rate of polymerization decreases, if a donor ligand on the Al component (C_2H_5) is replaced by an acceptor ligand. At the same time, however, the β-H abstraction is clearly increased indicating that acceptor ligands are favorable for this latter reaction (Henrici-Olivé and Olivé, 1971). Similar observations have been made with Ti/Al Ziegler type catalysts (Henrici-Olivé and Olivé, 1974).

Finally, it should be noted that the *"hydride character"* of metal hydrido complexes in aprotic solvents is markedly influenced by donor ligands, as shown by Richmond et al. (1982), by studying the reactions of such complexes with Lewis acids. For instance, HMn(CO)$_5$ reacts with BCl$_3$ giving a hydride bridged adduct (Eq. 5.44), whereas HMn(CO)$_4$(PPh$_3$) (one CO ligand replaced by the better σ donor, poorer π acceptor PPh$_3$) with BCl$_3$ results in complete hydride cleavage and the formation of the chloro derivative (Eq. 5.45):

$$HMn(CO)_5 + BCl_3 \rightleftharpoons Cl_3B-H-Mn(CO)_5 \tag{5.44}$$

$$HMn(CO)_4(PPh_3) \xrightarrow{\text{BCl}_3} ClMn(CO)_4(PPh_3). \tag{5.45}$$

Evidently, the donor ligand has made the hydride ligand more negative, and hence more susceptible to react with the Lewis acid.

In polar solvents, on the other hand, metal hydride complexes tend to be slightly acidic (see Table 2.2). Donor ligands affect the charge on the hydrogen in the same direction, making the hydrogen less positive (less acidic). Thus the dissociation constant of $Ru_4(CO)_{11}P(OCH_3)_3H_4$ (Nr. 11 in Table 2.2) is three orders of magnitude smaller than that or $Ru_4(CO)_{12}H_4$ (Nr. 10).

The ligand effects discussed in this Section are of great help for interpreting reactivity and selectivity in homogeneous catalysis in general, and in the homogeneous hydrogenation of CO in particular. They also can help to understand mechanistic features in heterogeneous CO hydrogenation, where isolated ligand effects cannot so easily be studied.

5.5.2 On Metal Surfaces

In a pure, clean metal particle, the only "ligands" surrounding a particular metal atom are, obviously, other metal atoms. Metal atoms at the surface have free coordination sites available, and metal atoms at crystal steps or kinks or other crystal defects even more so.

Theoretical MO calculations of Doyen and Ertl (1974) indicate the "ligand effect" of neighboring metal atoms. Based on a modified Hartree-Fock approximation, they estimated the metal$-$C bond energy for CO adsorbed on a metal surface, as a function of the number of metal atoms that contribute to this bond. Coupling to only one surface metal atom yields $\simeq 40\%$ of the full binding energy, whereas with 9 atoms 97.5% and with 25 atoms 100% is reached. It was concluded that an "ensemble" of about ten metal atoms contributes to the binding energy of an individual CO molecule. (Ligand influences from atoms several bonds away are documented in homogeneous coordination chemistry, see, e.g., Eqs. 5.38$-$5.41.) Recalling that the major factor in the metal$-$CO bond is the electron back-donation from the metal to the π^* orbital of CO (cf. Chap. 4), we can express the theoretical findings in terms of an electron donating ligand effect of the surrounding metal atoms. We have to conclude that, if a CO molecule is coordinatively bonded to one particular surface metal atom, the neighboring atoms will not be in a position to form an equally strong bond to other CO molecules. In fact, it is well known from adsorption/desorption experiments at increasing coverage (of CO or H_2) that a fraction of surface metal atoms with the strongest binding energy is occupied first, and only if all strongly binding sites are filled, other less binding sites are occupied (see e.g. Christmann and Ertl, 1976; Goodman et al., 1980); Peebles et al., 1983).

The importance of electron-donating-accepting processes for the binding between metal and substrate has been shown nicely by Russell et al. (1983), who measured the infrared spectrum of CO adsorbed on a Pt electrode, as a function of the applied potential. These in situ IR measurements were possible by an adaptation of the technique of polarization-modulation infrared reflection-absorption spectroscopy to the study of electrode surfaces. On increasing the anodic potential applied to the electrode in the range from 50 to 650 mV, ν_{CO} increased from 2077 to 2098 cm^{-1}. It is assumed that, as the Pt electrode becomes more positively polarized, electron density withdrawn from the metal is replaced by electron density formerly donated to the CO molecule, in the M−CO bond. A theoretical treatment (Ray and Anderson, 1982) of the problem confirms that it is just that: decreased back-donation into the π^* orbital of CO, with concomitant strengthening of the C−O bond and weakening of the metal−CO bond. From the calculations it follows that a potential change of the order of 1 V is required for a charge transfer of 0.1 electron. It is tempting (although probably limited by cost factors) to consider the application of this principle to catalytic processes, e.g. reducing the coordination ability of CO in favor of hydrogen (vide infra) by applying an anodic potential or, inversely, to provoke it, or even to dissociate adsorbed π-acceptor molecules, changing the sign of the applied potential.

Under the actual reaction conditions of catalytic CO hydrogenation, both CO and H_2 are available as ligands, and compete for free coordination sites. Carbon monoxide being the "better" ligand (CO stands right at the top of the spectrochemical series of ligands, see e.g. Henrici-Olivé and Olivé, 1977), it must be outweighed by hydrogen. Thus, most catalytic processes operate at $H_2/CO > 1$, generally $H_2/CO \geqq 2$. This is also expressed in the kinetic rate laws of these processes which are often of the type

$$r \sim p_{H_2}^n \, p_{CO}^m \qquad\qquad\qquad (5.46)$$

with $n \geqq 1$ and m often negative (see e.g. Falbe, 1977). It is evident that hydrogen does not have a chance to coordinate to the catalytic active site if CO occupies all coordination sites.

Examples of a deliberate change of the ligand environment of a heterogeneous metal (or metal oxide) catalyst are scarce. Actual catalysts are generally activated and regenerated under rather rough conditions (reductive treatment at high temperature), and not many possible ligands would survive that kind of a treatment. For metal oxide catalysts modification by other metal oxides has been reported. Taking into account the principle of electron-donation or -acceptance, one has to consider the "acidity" of metal oxides. MoO_3 is probably the most acidic metal oxide known; it easily dissolves even in dilute alkali. TiO_2, on the other hand, dissolves only poorly in concentrated alkali, and hence can be classified more on the basic side, i.e. acting as an electron donor towards most metal oxides. MoO_3 was used to modify the Phillips catalyst for the polymerization of ethylene, chromium oxide on silica (Henrici-Olivé and Olivé, 1973a). The Phillips catalyst usually generates high molecular weight polyethylene. The intention was to reduce the molecular weight towards the range of low molecular weight 1-alkenes, by increasing the β-H abstraction (chain transfer), applying acceptor ligands (cf. preceding Section). The catalysts were prepared by impregnating SiO_2 with aqueous solutions of CrO_3 and ammonium molybdate, oxidation at 500 °C in O_2 stream, and activation at 500 °C in CO stream. Table 5.3 shows some of the results. In the absence of Mo, high polymer is formed which, for the most part, is cracked under the reaction conditions, giving highly branched, liquid

Table 5.3. Influence of molybdenum additive on SiO_2-supported chromium oxide catalyst. Substrate: ethylene; $T = 300\,°C$; $[Cr] = 1 \times 10^{-3}$ g-atom per g catalyst (Henrici-Olivé and Olivé, 1973a).

Mo/Cr	Conversion[a] %	Product Distribution (%)				
		C_3	C_4	C_5	C_6	Polymer[b]
0	54.7	–	4.4	–	0.5	95.1
1	25.8	57.3	31.4	6.6	4.7	–
3	22.0	75.0	20.0	3.6	1.4	–

[a] Space velocity in [mol C_2H_4/(min × g-atom Cr)] = 0.65;
[b] This fraction includes highly branched cracked material.

oligomers of a molecular weight of 250–300, with negligeable unsaturation. Addition of the Mo species, on the other hand, *converts the Phillips catalyst into an oligomerization catalyst*. The growing chains leave the catalytic center after two or three propagation steps, as butene or hexene (β-H elimination). Only traces of octene and no higher oligomers were detected. The molybdenum additive not only drastically reduces the chain length, but also operates as a metathesis catalyst, producing oligomers with odd carbon numbers. (It should be noted that MoO_3 alone on SiO_2 is essentially inactive for the oligomerization of ethylene.) Since the insertion of CO (growth step) is favored by donor and not by acceptor ligands (preceding Section), it is not surprising that the overall rate of reaction is decreased by the Mo additive (Table 5.3). Magnetic measurements indicated that there is interaction between the two metals. The following structure was suggested for the active species:

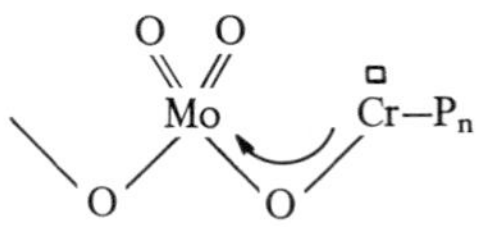

(with Pn = growing chain, and □ = vacant coordination site). The valency state of chromium in the active species is assumed to be Cr(II) in accordance with available evidence for homogeneous (Henrici-Olivé and Olivé, 1971) and heterogeneous (McDaniel, 1982) chromium catalysts.

The donor ligand TiO_2 has been applied in a similar way to the olefin metathesis catalyst molybdenum oxide on alumina (Henrici-Olivé and Olivé, 1973b). The result was a considerable increase of the rate of the metathesis reaction, e.g.:

$$2\;CH_3CH{=}CH_2 \rightleftarrows CH_2{=}CH_2 + CH_3CH{=}CHCH_3.$$

Here, again, magnetic data (electron paramagnetic resonance) indicate that the environment of the Mo centers changes if TiO_2 is introduced. The structure

was suggested for the active site. Cr/silica polymerization catalysts are also rendered much more active by the incorporation of titania (Pullukat et al., 1981; McDaniel et al., 1983).

Alloying a catalytically active metal with an inactive one can be considered as a ligand effect. Nickel/copper alloys have been widely investigated with regard to their activity for CO hydrogenation (Araki and Ponec, 1976; Barneveld and Ponec, 1978). It was found that alloying reduces the activity of Ni more than it could be expected from mere dilution, and it was concluded that undisturbed "Ni ensembles" are necessary for catalytic activity. Burch (1981) suggested that this is an electronic effect. Although the position of the Ni $3d$ band is hardly affected by adding Cu, the band width is strongly affected, becoming narrower with increasing number of Cu next neighbors. The result is that the d band slips below the Fermi level, i.e. it becomes filled. As a consequence the Ni becomes less "Ni-like". An estimated ensemble of 12 Ni atoms would be required to ensure that the central Ni atom has its "full Ni character".

Finally, for finely divided metal catalysts on a support, the support material may act as a ligand, and this effect has found widespread attention in the past years. Although metal-support interaction will be discussed in greater detail in the next Chapter (Sect. 6.2), one illustrative example will be presented here. The magnetic properties of chromium catalysts depend on the support. Table 5.4 shows the effective magnetic moment, μ_{eff}, for Cr/SiO_2 and Cr/Al_2O_3, as determined by measurements of the magnetic susceptibility (Henrici-Olivé and Olivé, 1983). The samples were prepared impregnating the support with aqueous CrO_3, followed by oxidation (8 h, O_2, 500 °C) and

Table 5.4. Magnetic moment μ_{eff} of reduced chromium oxide catalysts. [Cr] = 1.1×10^{-4} g-atom/g catalyst (Henrici-Olivé and Olivé, 1983).

Support	Reduction time[a] (h)	μ_{eff} (BM)
SiO_2	1	2.43
SiO_2	3	2.65
SiO_2	7	2.50
Al_2O_3	3	3.73
Al_2O_3	7	4.60

[a] 500 °C, CO stream.

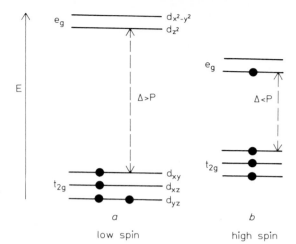

Fig. 5.4. Splitting of the d levels of Cr(II) ($3d^4$) in an octahedral complex; a) strong ligand field, energy gap Δ larger than the spin pairing energy P (e.g. Cr(II)/SiO$_2$); b) weak ligand field (e.g. Cr(II)/Al$_2$O$_3$)

reduction (CO, 500 °C, variable times). On the SiO$_2$ support, reduction leads to a μ_{eff} value roughly corresponding to two unpaired electrons per Cr atom (2.83 BM), and does not proceed any further even after 7 h of reduction. On Al$_2$O$_3$ support, however, the number of unpaired electrons increases further, tending towards 4.9 BM, value corresponding to four unpaired electrons per Cr atom. (This clear-cut difference is only observed at low chromium concentration; at higher concentration reduction is incomplete, and a μ_{eff} value corresponding to a mixture of valency states is observed; Henrici-Olivé and Olivé, 1973 a.) Cr(II) ($3d^4$) complexes are octahedral with few exceptions (Cotton and Wilkinson, 1980). In this symmetry there are two possible configurations for the four d electrons, depending on the energy gap Δ between the t$_{2g}$ and e$_g$ orbitals (see e.g. Henrici-Olivé and Olivé, 1977). In a strong ligand field Δ is large, and the energetically more favorable configuration is that having all four electrons in the low lying t$_{2g}$ level, despite the energy required for spin pairing (see Fig. 5.4a). In a weak ligand field Δ is small, and the situation with four unpaired electrons in four orbitals is more favorable (Fig. 5.4b). The data of Table 5.4 then indicate a strong ligand field in the case of SiO$_2$ support, a weak ligand field for Al$_2$O$_3$. Presumably both supports interact with the chromium via oxygen, the difference in ligand field might be caused by different Cr−O distances. One possible explanation could be chemical bond formation in the SiO$_2$ case (as suggested by McDaniel, 1982), and non bonding interaction with Al$_2$O$_3$. The data in Table 5.4 certainly show that the support is not just a carrier of the catalytic active species but can have strong interaction with the latter. A considerable influence on the catalytic activity can be the consequence (see Sect. 6.2).

5.6 Conclusion

It appears that much of the wealth of knowledge accumulated during the past decades in the field of homogeneous catalysis may be carried over to heterogeneous catalysis. The most important key reactions do take place at metal surfaces, for instance oxidative addition/reductive elimination (during adsorption and thermal desorption of hydrogen), or CO insertion (in the formation of formyl surface species), and ligand influences appear to be governed by the same principles. Nevertheless, there are differences between homogeneous and heterogeneous processes. Thus, for instance, no homogeneous (soluble) catalyst has been found to make large hydrocarbon molecules from CO and H_2; on the other hand no hydroformylation has been observed in heterogeneous solid/gas systems (where "heterogenized" catalysts have been used for hydroformylation, they were applied in dispersion, i.e. with a liquid medium available). Some of the analogies and discrepancies between homogeneous and heterogeneous catalysts will be further discussed in the next Chapter.

5.7 References

Abis L, Sen A, Halpern J (1978) J Amer Chem Soc 100:2915

Alt HG, Eichner ME (1982) Angew Chem 94:68; Angew Chem Int Ed Engl 21:78

Anderson GK, Cross RJ (1980) J Chem Soc Dalton, 712 Ank NT (1976) Tetrahedron Lett 155

Araki M, Ponec V (1976) J Catal 44:439

van Barneveld WAA, Ponec V (1978) J Catal 51:426

Berke H, Hoffmann R (1978) J Amer Chem Soc 100:7224

Bradley JS (1979) J Amer Chem Soc 101:7421

Brookhart M, Tucker JR, Husk GR (1981) J Amer Chem Soc 103:979

Brown MP, Fisher JR, Mills AJ, Puddephatt RJ, Thomson M (1980) Inorg Chim Acta 44:L271

Burch R (1981) J Chem Soc Chem Commun 845

Butts SB, Strauss SH, Holt EM, Stimson RE, Alcock NW, Shriver DF (1980) J Amer Chem Soc 102:5093

Byers BH, Brown TL (1977) J Organometal Chem 127:181

Calderazzo F (1977) Angew Chem 89:305; Angew Chem Int Ed Engl 16:299

Calderazzo F, Cotton FA (1962a) Inorg Chem 1:30

Calderazzo F, Cotton FA (1962b) Abstracts of the Internat. Conference on Coordination Chemistry, Stockholm, paper 6H7

Calvert RB, Shapley JR (1977) J Amer Chem Soc 99:2533

Calvert RB, Shapley JR (1978) J Amer Chem Soc 100:7726

Canestrari M, Green MLH (1982) J Chem Soc Dalton Trans 1789

Casey CP (1979) CHEMTECH (june), 380

Casey CP, Bunnell CA, Calabrese JC (1976) J Amer Chem Soc 98:1166

Cheng C-H, Spivack BD, Eisenberg R (1977) J Amer Chem Soc 99:3003

Christmann K, Ertl G (1976) Surface Sci 60: 356

Churchill MR, Wasserman HJ, Turner HW, Schrock RR (1982) J Amer Chem Soc 104:1710

Collman JP, Hegedus LS (1980) Principles and Applications of Organotransition Metal Chemistry. University Science Books, Mill Valley, California

Collman JP, Finke RG, Cawse JN, Bramman JI (1978) J Amer Chem Soc 100:4766
Connor JA (1977) Topics Curr Chem 71:71
Cotton FA, Wilkinson G (1980) Advanced Inorganic Chemistry. Interscience, 4th ed
Cramer R (1965) J Amer Chem Soc 87:4717
Cramer R (1972) J Amer Chem Soc 94:5681
Cross RJ, Gemmill J (1981) J Chem Soc Dalton Trans 2317
Dawoodi Z, Green MLH, Mtetwa VSB, Prout K (1982a) J Chem Soc Chem Commun 802
Dawoodi Z, Green MLH, Mtetwa VSB, Prout K (1982b) J Chem Soc Chem Commun 1410
Doyen G, Ertl G (1974) Surface Sci 43:197
Egger KW (1969) J Chem Kinet 1:459
Evans J, Schwartz J, Urguhart PW (1974) J Organometal Chem 81:C37
Evitt ER, Bergman G (1979) J Amer Chem Soc 101:3973
Fachinetti G, Floriani C, Stoeckli-Evans H (1977) J Chem Soc Dalton 2297
Falbe J (1977) Chemierohstoffe aus Kohle. Georg Thieme Verlag, Stuttgart
Flood FC, Jensen JE, Statler JA (1981) J Amer Chem Soc 103:4410
Foley P, DiCosimo R, Whitesides GM (1980) J Amer Chem Soc 102:6713
Gillie A, Stille JK (1980) J Amer Chem Soc 102:4393
Gladysz JA, Selover JC (1978) Tetrahedron Lett 319
Goddard RJ, Hoffmann R, Jemmis ED (1980) J Amer Chem Soc 102:7667
Goodman DW, Yates JT, Madey TE (1980) Surface Sci 93:L135
Green MLH (1978) Pure Appl Chem 50:27
Halpern J (1982) Acc Chem Res 15:332
Halpern J, Chock PB (1966) J Amer Chem Soc 88:3511
Hayes JC, Pearson GDN, Cooper NJ (1981) J Amer Chem Soc 103:4648
Henrici-Olivé G, Olivé S (1969) Adv Polymer Sci 6:421
Henrici-Olivé G, Olivé S (1971) Angew Chem 83:782; Angew Chem Int Ed Engl 10:776
Henrici-Olivé G, Olivé S (1973a) Angew Chem 85:827; Angew Chem Int Ed Engl 12:754
Henrici-Olivé G, Olivé S (1973b) Angew Chem 85:148; Angew Chem Int Ed Engl 12:153
Henrici-Olivé G, Olivé S (1974) Adv Polymer Sci 15:1
Henrici-Olivé G, Olivé S (1977) Coordination and Catalysis, Chap. 7, Verlag Chemie,
    Weinheim, New York
Henrici-Olivé G, Olivé S (1977/78) J Molecular Catal 3:443
Henrici-Olivé G, Olivé S (1983) unpublished results
Hoffmann R (1981) Front Chem Plenary Keynote Lect., IUPAC Congr. 28th 1981, published
    1982 (Laidler KJ, ed), Pergamon, Oxford, UK
Isobe K, Andrews DG, Mann BE, Maitlis PM (1981) J Chem Soc Chem Commun 809
James BR (1973) Homogeneous Hydrogenation. Wiley, New York
Jones WD, Bergman RG (1979) J Amer Chem Soc 101:5447
Kampe CE, Boag NM, Kaesz HD (1983) J Amer Chem Soc 105:2896
Klein HF, Karsch HH (1976) Chem Ber 109:2524
Lalage D, Brown S, Connor JA, Skinner HA (1974) J Organometal Chem 81:403
Lin YC, Calabrese JC, Wreford SS (1983) J Amer Chem Soc 105:1679
McDaniel MP (1982) J Catal 76:17
McDaniel MP, Welch MB, Dreiling MJ (1983) J Catal 82:118
Magnuson RH, Meirowitz R, Zulu SJ, Giering WP (1983) Organometallics 2:460
Michman M, Zeiss HH (1968) J Organometal Chem 13:23; 15:139
Milstein D (1982) J Amer Chem Soc 104:5227
Morrison ED, Steinmetz GR, Geoffroy GL, Fultz WC, Rheingold AL (1983) J Amer Chem
    Soc 105:4104
Nappa MJ, Santi R, Diefenbach SP, Halpern J (1982) J Amer Chem Soc 104:619
Noack K, Calderazzo F (1967) J Organometal Chem 10:101
Noell JO, Hay PJ (1982) J Amer Chem Soc 104:4578
Norton JR (1979) Acc Chem Res 12:139
O'Neal HE, Benson SW (1967) J Phys Chem 71:2903
Peebles DE, Creighton JR, Belton DN, White JM (1983) J Catal 80:482
Pu LS, Yamamoto A (1974) J Chem Soc Chem Commun 9
Pullukat TJ, Hoff RE, Shida M (1981) J Appl Polymer Sci 26:2927

Ray NK, Anderson AB (1982) J Phys Chem 86:4851
Richmond TG, Basolo F, Shriver DF (1982) Organometallics 1:1624
Roth JA, Orchin M (1979) J Organometal Chem 172:C27
Russell JW, Severson M, Scanlon K, Overend J, Bewick A (1983) J Phys Chem 87:293
Sakaki S, Kitaura K, Morokuma K, Ohkubo K (1983) J Amer Chem Soc 105:2280
Sen A, Lai T-W (1982) J Amer Chem Soc 104:3520
Shustorovich E, Baetzold RC, Muetterties EL (1983) J Phys Chem 87:1100
Soto J, Steigerwald ML, Grubbs RH (1982) J Amer Chem Soc 104:4479
Thorn DL (1982) Organometallics 1:879
Thorn DL, Tulip TH (1981) J Amer Chem Soc 103:5984
Threlkel RS, Bercaw JE (1981) J Amer Chem Soc 103:2650
Tolman CA, Meakin PZ, Lindner DL, Jesson JP (1974) J Amer Chem Soc 96:2762
Tulip TH, Thorn DL (1981) J Amer Chem Soc 103:2448
Vaska L, DiLuzio JW (1962) J Amer Chem Soc 84:679
Vaska L, Werneke MF (1971) Trans NY Acad Sci 31:70
Vaughn GD, Gladysz JA (1981) J Amer Chem Soc 103:5608
Wayland BB, Woods BA, Minda VM (1982) J Chem Soc Chem Commun 634
Wax MJ, Bergman RG (1981) J Amer Chem Soc 103:7028
Westheimer FH (1961) Chem Rev 61:265
Zumdahl SS, Drago RS (1968) J Amer Chem Soc 90:6669

6 Catalysts and Supports

6.1 Molecular Complexes and Metal Surfaces – Analogies and Differences

Catalysts for the hydrogenation of carbon monoxide are found among *molecular transition metal carbonyl complexes* as well as *metal surfaces*. Knowing the close analogies between the two types of metal centers, as they have been emphasized in the preceding Chapters, this cannot be a surprise. However, with a few exceptions, the reaction products obtained from $CO + H_2$ are quite different. Thus, for instance, glycol and other polyalcohols have only been produced by *homogeneous* catalysts (i.e. soluble, molecular complexes), whereas long chain hydrocarbons are formed exclusively by *heterogeneous* catalysts (i.e. metal surfaces, or small metal particles on a support). The question arises as to what is at the origin of this type of specialization.

Several suggestions have been forwarded in the course of the years. The "ensemble" hypothesis assumes that the *presence of groups* (ensembles) of similar metal atoms on adjacent metal sites is required for "demanding" catalytic reactions, as for instance the chain growth in the Fischer-Tropsch synthesis (see e.g. Ponec and Sachtler, 1972; Doyen and Ertl, 1974; Araki and Ponec, 1976). These ideas, together with the advent, in the course of the past decade, of a new branch of coordination chemistry, namely that of large metal carbonyl clusters (see Sect. 3.2), have spurred great hopes and a considerable research effort with regard to the catalytic prospects of these clusters. The expectation was that these soluble clusters might close the gap between the mononuclear homogeneous catalysts and the heterogeneous metal particles. In fact the similarities between metal surfaces and clusters are often amazing:

the metal-metal distances are essentially the same;
the metal−CO bond strengths are comparable;
ligand mobility is characterized by the same activation energy (for reviews on cluster/surface analogies see Basset and Ugo, 1977; Muetterties et al., 1979).

Nevertheless, the results on the catalytic front were very disappointing, at least as far as CO hydrogenation is concerned (Muetterties and Krause, 1983, and references therein). In some cases solutions of carbonyl clusters, e.g. $Os_3(CO)_{12}$ or $Ir_4(CO)_{12}$, react with H_2 and CO under mild conditions to produce CH_4 and some light hydrocarbons, but the rates are so low that it is unclear whether the reactions are truly catalytic or not.

The catalytic conversion of $CO + H_2$ to hydrocarbons with the cluster compound $Ru_3(CO)_{12}$ was traced back to ruthenium metal, generated by the thermal decomposition of the cluster.

The system $Ir_4(CO)_{12}$ in molten $AlCl_3$-$NaCl$ was reported to produce methane, ethane, propane and higher branched alkanes (Demitras and Muetterties, 1977).

Collman et al. (1983), studying the mechanistic aspects of this interesting reaction, found experimental evidence showing that it involves the homogeneous reduction of CO to chloromethane, followed by homologation and/or hydrogenation reactions leading to the hydrocarbon products. Hence, this reaction does not appear to be a homogeneous analog of the classical Fischer-Tropsch synthesis. Moreover, in this and similar reactions (using BBr_3 or BCl_3 as a reaction medium), the "solvents" $AlCl_3$, BBr_3 or BCl_3 were consumed, because the oxygen atoms of the CO molecules which were converted into hydrocarbons or alkyl halides, ended up bound to aluminum or boron (Muetterties and Krause, 1983).

Another relevant observation was reported by Halpern (1982), who studied oxidative addition and reductive elimination reactions on binuclear cluster compounds. He found that either the binuclear species broke apart, the active species being mononuclear, or the reaction took place only on one of the metal atoms.

Taken together, the data reported thus far appear to indicate that the presence of a geometric array of metal atoms in adjacent positions is not the clue — at least not the only one — to the different catalytic behavior of metal surfaces as compared with homogeneous systems. The major reason why large carbonyl clusters are not able to fill that gap has been mentioned in Sect. 3.2: most cluster carbonyls are coordinatively saturated compounds, that means *they have all low lying molecular orbitals filled, and display generally a large energy difference between the highest occupied and the lowest unoccupied orbitals, what contributes to their inactivity.* Surface metal atoms, on the other hand, have inherently an incomplete coordination sphere.

Another line of thoughts concerns itself with the number of metal atoms needed for a metal particle to show "bulk like" properties, i.e. with regard to *d*-band width; estimates vary from about ten to several hundreds of atoms, depending on the model calculation, or on the experimental method applied (Baetzold et al., 1980). But most probably the question where bulk-like properties begin — interesting as it may be for the surface scientist — is not all that relevant for catalytic activity. It appears to be more the particular situation of surface atoms that cannot be offered by the cluster compounds. Such surface atoms (especially at steps, kinks or other surface defects) may have just the right combination of available coordination sites and *d*-electron energy. Johnson et al. (1978) compared the local electronic structure of a surface metal atom with that of the metal in a coordinatively unsaturated metal complex (e.g. $Pt(PH_3)_2$). The results of these calculations are shown in Fig. 6.1.

The mononuclear complex (Fig. 6.1 a) is characterized by a strong bonding interaction between the ligand lone-pair *p*-like orbitals and the metal *d* orbitals along the metal-ligand

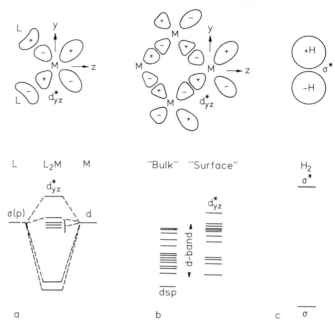

Fig. 6.1 a – c. Comparison of the electronic structures associated with coordinatively unsaturated transition metal species: **a)** in a L_2M complex (e.g. $L=Pt(PH_3)_2$); **b)** at the surface of a transition metal particle. For comparison: **c)** the bonding and antibonding orbitals of H_2 (according to Johnson et al., 1978). Reproduced by permission of North-Holland Physics Publishing

direction (in the yz plane for coordinate system shown). The antibonding component of this interaction pushes the metal d_{yz}^* orbital to a high energy position, as compared with the nonbonding d_{zy}, d_{xz} and $d_{x^2-y^2}$ orbitals. To approach the situation of the surface metal atom, the two ligands have been replaced by successively larger groups of metal atoms (M_x; $x = 1, 2, 3, \ldots, n$). The resulting manifold of bonding, nonbonding and antibonding d-orbitals (the d-band), by virtue of the component orbital symmetries and charge distributions, can be partitioned into "bulk-like" states (delocalized throughout the particle) and "surface-like" states (localized at the coordinatively unsaturated surface atom). In close analogy with Fig. 6.1 a, Fig. 6.1 b indicates that the direct interaction of the d-orbitals of the surface metal atom with the d-orbitals on neighboring metal sites (viewed as "ligands") leads to the splitting off in energy of a strongly antibonding d_{yz}^* orbital from the top of the main d-orbital manifold. (For other cluster geometries the most strongly antibonding d-orbital can have a different local symmetry, e.g. d_{z^2}, but the same principles would apply.) The d_{yz}^* orbital, if occupied, is in perfect conditions, energetically as well as with regard to symmetry, to interact with, for instance, the antibonding σ^* molecular orbital of H_2 (Fig. 6.1 c), as the starting move towards oxidative addition (and activation) of the hydrogen molecule.

Considerations of this type, though instructive for the interpretation of catalytic activity of surface metal atoms, still fall short of answering the question why soluble mononuclear, coordinatively unsaturated complexes are not able to catalyze the same reactions as heterogeneous catalysts (e.g. Fischer-Tropsch synthesis).

We may add two more points of view. One distinctive characteristic of the heterogeneous systems is their ability to open up the C–O bond, either forming metal carbides (cf. Chap. 7) or by dehydration of intermediates (cf. Chap. 9). The main products of the homogeneous reactions, on the other hand, contain undissociated C—O bonds: polyalcohols, glycolaldehyde from CO and H_2, aldehydes from alkenes, CO and H_2 (Chap. 10).

The other distinctive feature is the mobility and availability of metal-hydrides in the homogeneous reactions which can lead to intermolecular reductive elimination of intermediates in a way not possible on solid catalysts (cf. Sect. 5.2.2).

6.2 Supported Metal Catalysts

6.2.1 Metals and Supports Used for CO Hydrogenation

Heterogeneous catalysis requires, in general, that the catalytic metal is present *in a highly dispersed form* so as to maximize the surface area available for a given mass of metal. (For a review on the structure of metallic catalysts see Anderson, 1975.) In most instances, this is best achieved by distributing the metal particles on a support. The most frequently used supports (or carriers) are

SiO_2 and Al_2O_3, but also TiO_2, MgO, ZnO;

other porous oxides are also suitable in certain cases. The supported metal particles are stable towards agglomeration, as well as accessible to the reacting gases. The ratio metal/support varies in a wide range, from a few % to > 50%.

Several methods are available to prepare supported catalysts. The most important are:

a) *Impregnation.* Solutions of metal nitrates, oxalates, or other easily de-composable salts or complexes (in particular carbonyl complexes) of the metal are used to impregnate the support (generally in form of porous granules of the desired size). After drying, the catalyst is oxidized to eliminate anions and/or ligands (usually at 500 °C or more, O_2 stream), followed by reduction to metal (500 °C or more, H_2 or H_2/CO stream). A sideline of this procedure consists in "impregnating" the degassed carrier with gaseous metal carbonyls (e.g. $Fe(CO)_5$), followed by thermal decom-position of the complex.

b) *Precipitation.* The hydroxide or hydroxycarbonate of the metal is pre-cipitated from a metal salt solution in the presence of the support, followed by washing, drying, oxidation and reduction. Coprecipitation of metal and support has also been used.

c) *Co-melting.* Support and metal oxide or carbonate are melted together in an electric oven, followed by crushing of the cake.

The metals used in the catalytic hydrogenation of carbon monoxide are mainly the group VIII metals of the first and second transition series, copper (most probably Cu(I) d^{10}), and to a lesser degree Pt. Apparently, these metals dispose of just the right combination of high energy, filled d orbitals (cf. Fig. 6.1b) for interacting with the σ^* orbital of the H_2, thus activating the latter for oxidative addition, and for providing a destabilization of coordinatively bonded CO by a π-back donation. Moreover they make available empty (or half filled) d orbitals for the corresponding σ-bonds (i.e. $M-H$, $M-CO$, as well as bonds between the metal and reaction intermediates).

The reaction products of the CO hydrogenation are manifold, ranging from methane, methanol, long chain hydrocarbons and primary alcohols to glycol and polyalcohols.

Certainly not every metal of those mentioned above is able to catalyze the formation of each of those compounds; rather there is a certain "specialization". However, not only the metal itself, but also support, temperature, pressure, H_2/CO ratio, and other variables have an influence on the outcome of any particular CO hydrogenation. Thus, the following "assignment" of metals to special tasks can only be a very rough guideline.

Nickel, palladium and *platinum* tend to produce methane;
iron, cobalt, and *ruthenium* form higher hydrocarbons and monoalcohols;
copper (especially with ZnO) catalyzes the production of methanol. (In homogeneous catalysis, Rh and Ru are the specialists for the formation of methanol, glycol and higher polyalcohols.)
Osmium and *iridium* react very sluggishly with CO/H_2; they have not found use as catalysts, however iridium has proven to be very useful in modeling certain suggested steps in complex CO hydrogenation mechanisms.

6.2.2 Metal-Support Interactions

It has been known since the old times of the Fischer-Tropsch synthesis in the thirties that the support is not only a physical carrier for the metallic catalyst, but that it can have influence on *rate* and *selectivity* of the CO hydrogenation (Storch et al., 1951). The investigation of this influence has recently received new impulse, spurred by the observation of a particularly strong metal-support interaction (abbreviated SMSI) in the case of *titania supports* (Tauster et al., 1978).

SMSI is observed if Ru, Rh, Pd, Os, Ir, Pt (Tauster et al., 1981) or Ni (Vannice and Garten, 1979) is deposited on TiO_2, and the final reductive treatment of the catalyst is carried out at 500 °C or higher, with *hydrogen*.

The effect manifests itself by a remarkable, and sometimes nearly complete suppression of the hydrogen and carbon monoxide adsorption (see Table 6.1). An investigation of other oxides showed that only Nb_2O_5 had a similar effect, and it was concluded that the relative ease of reduction of TiO_2 and Nb_2O_5 may play a role (Tauster et al., 1981). In fact, electron diffraction indicated the formation of a lower oxide, Ti_4O_7. Electron microscopic studies revealed a

Table 6.1. Hydrogen and carbon monoxide adsorption on TiO_2 supported metals, after reductive treatment at different temperature T_R (according to Tauster et al., 1981).

Metal[a]	Hydrogen adsorption[b]		CO adsorption[c]	
	$T_R = 200\,°C$	$T_R = 500\,°C$	$T_R = 200\,°C$	$T_R = 500\,°C$
Ru	0.23	0.06	0.64	0.11
Rh	0.71	0.01	1.15	0.02
Pd	0.93	0.05	0.53	0.02
Os	0.21	0.11	–	–
Ir	1.60	0.00	1.19	0.00
Pt	0.88	0.00	0.65	0.03

[a] 2 wt%;
[b] ratio of hydrogen atoms adsorbed to total metal atoms;
[c] ratio of CO molecules adsorbed to total metal atoms.

much smaller mean particle size after high temperature treatment (SMSI state) than after low temperature treatment; this difference was not found for other carriers such as SiO_2 or Al_2O_3. High temperature oxidation destroys the SMSI state, as evidenced by a return to normal adsorption properties and increased particle size. The effect was found to be reversible, by repetition of the high temperature treatment with H_2. It was assumed that the metal catalyst provides H atoms by dissociation of H_2; the H atoms migrate to the support and reduce it to a lower oxidation state, Ti_4O_7. The interaction of the latter with the metal is sufficiently strong to overcome the cohesive forces in the metal aggregates, partitioning them into smaller particles. Hydrogen atom participation in the process was corroborated by the observation that a Ag/TiO_2 system did not show SMSI behavior, but introduction of Pt onto the Ag/TiO_2 system, followed by a reductive cycle, led to the SMSI state, including the partitioning of the Ag particles (Baker et al., 1983). A theoretical treatment of the SMSI phenomenon (Horsley, 1979) suggests transfer of electron density from subjacent carrier cations to the metal atoms, resulting in a strong bond between negatively charged small metal particles and the associated cations in the support surface.

Santos et al. (1983) observed SMSI behavior in the system Fe/TiO_2, at least as far as suppression of CO adsorption is concerned. Even relatively large particles (up to 20 nm) showed the effect; their Mössbauer parameters, however, were identical with those of normal bulk metallic iron. Hence, the metal-support interaction appears to be restricted to the surface of the metallic particles. It was concluded that titanium species are transported from the support to the surface of the metallic particles in the course of the catalyst preparation and/or reduction at high temperature.

The study of these effects, involving the almost total inhibition of CO and H_2 adsorption, may appear counterproductive at first sight. However, it was found that Ni/TiO_2 catalysts are an order of magnitude more active for the reduction of CO to methane, than Ni/Al_2O_3 or Ni/SiO_2 with comparable over-all Ni content (Vannice and Garten, 1979). For supported Pt catalysts, an in-crease in activity with variation of the support was found in the series

$$Pt/SiO_2 : Pt/Al_2O_3 : Pt/TiO_2 = 1 : 10 : 100$$

(Vannice and Twu, 1983). The relation between suppression of adsorption and increase of activity is certainly enigmatic, and the exact nature of the effect is not known (Vannice and Twu, 1983). We suggest that a possible explanation may go along the following lines:
The strong interaction may suppress the adsorption on plain crystal surfaces, which are not catalytically active anyway, but not on steps and kinks (see

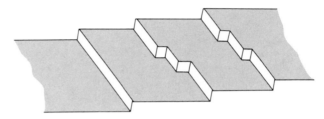

Fig. 6.2. Schematic representation of plain crystal surfaces, steps and kinks (according to Anderson, 1975). Reproduced by permission of Academic Press

Fig. 6.2), where the active sites are supposed to be located. These special locations, which feature metal atoms with a larger number of free coordination sites, are certainly more abundant in small particles than in large ones (at the same overall metal charge).

In fact, Vannice and Twu (1983) observed the same enhancement of methane formation from $CO + H_2$ with a Pt/TiO_2 catalyst reduced at 200 °C and one reduced at 500 °C, provided a high degree of dispersion of the metal was induced in the former by a special treatment (10 h purge with helium at room temperature, followed by heating to 200 °C before H_2 was substituted for He). Interestingly, the catalyst reduced at the lower temperature did not show SMSI behavior with regard to CO and H_2, both gases being adsorbed normally. This indicates that *the concentration of special active sites is in fact the important part of SMSI, and not the influence on gas adsorption on the overall surface.* Another important point is probably the electron donor ability of TiO_2, as evidenced by Mo/TiO_2 catalysts (cf. Sect. 5.5.2) and rationalized by the theoretical work of Horsley (1979). Electron donation to the metal should have a positive effect on CO insertion and oxidative addition reactions, both of which play an important role in CO hydrogenation ("ligand effect", see Sect. 5.5.1).

It may be added that SMSI is not at all favorable for all types of catalyzed reactions. Thus, Burch and Flambard (1981) found that the formation of methane was improved by a factor of 20 with Ni/TiO$_2$ as compared with Ni/SiO$_2$, while the hydrogenolysis of n-hexane took place at roughly the same rate with both catalysts. On the other hand, Santos et al. (1983) found that their SMSI Fe/TiO$_2$ catalyst was detrimental for the ammonia synthesis. Together, these investigations appear to indicate that some reaction involving CO, presumably the CO insertion, benefits most from the donor ability of TiO$_2$.

The donor acceptor relationship between support and metal is not restricted to the titania support, although it appears to be particularly drastic there. A direct measurement of this kind of effect on adsorbed CO has been made by Blackmond and Goodwin (1981), with ruthenium on zeolite support. Zeolites with increasing Si/Al ratio (increasing acidity, i.e. increasing electron acceptor capacity) were used. CO was adsorbed onto the supported Ru catalysts, and the infrared spectra were recorded. A marked increase of the v_{CO} with increasing acidity showed clearly the ligand influence of the support: decreasing electron density at the metal reduced the back donation metal$-$CO, hence reduced the electron density in the CO π^* orbital, i.e strengthened the C$-$O bond. This effect is accompanied by a decrease in catalytic activity of the Ru/zeolite catalyst for methane formation with increasing Si/Al ratio.

6.2.3 Steric Constraints in Zeolites

Apart from their possible electronic effect (preceding Section), zeolites have an interesting *sieve function*. They are crystalline aluminosilicates, the primary building blocks of which are tetrahedra consisting of either silicon or aluminum ions surrounded by four oxygen anions. These tetrahedra combine, linking silicon and aluminum ions together through oxygen bridges, to yield highly ordered three dimensional frameworks. There are 34 known natural and about 120 synthetic zeolite structures (Whan, 1981). All have a three-

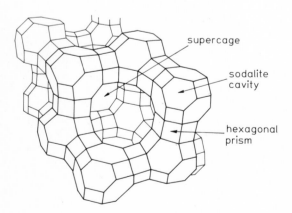

supercage

sodalite cavity

hexagonal prism

Fig. 6.3. Framework of a faujasite type zeolite. Supercage: 1.25 nm diameter, 0.75 nm entrance opening; sodalite cavity: 0.65 nm diameter, 0.22 nm entrance opening; hexagonal prism: 0.22 nm diameter (Ballivet-Tkatchenko et al., 1979). Reproduced by permission of Plenum Publ. Corp.

dimensional network of channels, or linked cavities. Figure 6.3 shows the framework of one of them, the faujasite type zeolite. In this particular case, the "super cage" has a diameter of 1.25 nm; the opening consisting of a 12-membered oxygen ring has a diameter of 0.75 nm (Ballivet-Tkatchenko et al., 1979). Whereas the silicon-oxygen tetrahedra building blocks are electrically neutral, each aluminum based tetrahedron has a negative charge, balanced by a cation. Formally, this may be written:

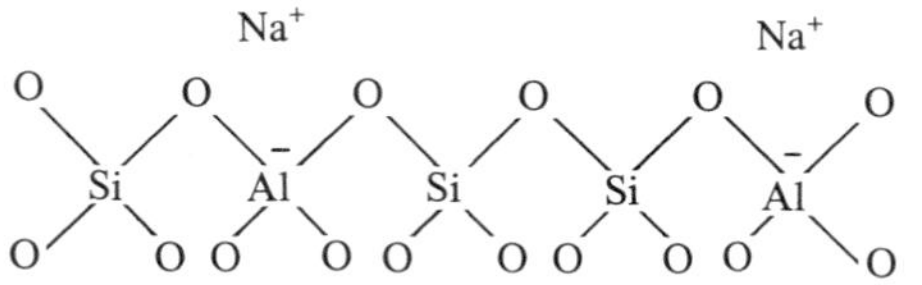

$$
\tag{6.1}
$$

Transition metal cations can be introduced by ion exchange. The zeolites are used as "shape selective" catalysts since the early sixties for a variety of reactions, whereby the product distribution depends on the ease with which reactant molecules diffuse through the zeolite channels, and on the residence time during which they are held in the vicinity of a catalytically active site. Ion exchange, cracking, hydrocracking, as well as "selectoforming" are among the most important industrial uses.

For the present it suffices to give several examples of the amazing molecular shape selectivity of zeolite catalysts, as reported by Weisz (1980). The dehydration of n-butanol on a zeolite catalyst is rapid, whereas isobutanol barely reacts; from mixtures of linear and branched hydrocarbons (e.g. n-butane/isobutane) the selective catalyzed combustion of the linear hydrocarbons could be achieved with Pt containing zeolites; selective hydrogenation of ethylene in a mixture of ethylene and propylene was attained.

Examples of the use of zeolite catalysts in the hydrogenation of CO will be discussed in later Chapters.

6.3 Conclusions

In this Chapter we have tried to give a general view of the various types of catalysts used for the hydrogenation of CO. In the following Chapters, which deal more specifically with the various mechanisms of the carbon monoxide hydrogenation such as methanation, methanol formation, Fischer-Tropsch synthesis, etc., the problem of the catalytically active species will be met again and again.

6.4 References

Anderson JR (1975) Structure of Metallic Catalysts. Academic Press, London, New York, San Francisco

Araki M, Ponec V (1976) Catal 44:439

Baetzold RC, Mason MG, Hamilton JF (1980) J Chem Phys 72:366

Baker RTK, Prestridge EB, Murrell LL (1983) J Catal 79:348

Ballivet-Tkatchenko D, Coudurier G, Mozzanega H, Tkatchenko I (1979) in: Tsutsui M (ed) Fundamental Research in Homogeneous Catalysis. Vol 3, p 257, Plenum Publ Corp

Basset JM, Ugo R (1977) in, Ugo R (ed) Aspects of Homogeneous Catalysis. Vol 3, Chapter 2, D. Reidel Publ. Co., Dordrecht, Holland

Blackmond DG, Goodwin JG (1981) J Chem Soc Chem Commun 125

Burch R, Flambard AR (1981) J Chem Soc Chem Commun 123

Collman JP, Brauman JI, Tustin G, Wann GS (1983) J Amer Chem Soc 105:3913

Demitras GC, Muetterties EL (1977) J Amer Chem Soc 99:2796

Doyen G, Ertl G (1974) Surface Sci 43:197

Halpern J (1982) Inorganica Acta 62:31

Horsley JA (1979) J Amer Chem Soc 101:2870

Johnson KH, Balasz AC, Kolari HJ (1978) Surface Sci 72:733

Muetterties EL, Krause MJ (1983) Angew Chem 95:135; Angew Chem Int Ed Engl 22:135

Muetterties EL, Rhodin TN, Band E, Brucker CF, Pretzer WR (1979) Chem Rev 79:91

Ponec V, Sachtler WMH (1972) J Catal 24:250

Santos J, Phillips J, Dumesic JA (1983) J Catal 81:147

Storch HH, Golumbic N, Anderson RB (1951) The Fischer-Tropsch and Related Syntheses. Wiley, New York

Tauster SJ, Fung SC, Garten RL (1978) J Amer Chem Soc 100:170

Tauster SJ, Fung SC, Baker RTK, Horsley JA (1981) Science 211:1121

Vannice MA, Garten RL (1979) J Catal 56:236

Vannice MA, Twu CC (1983) J Catal 82:213

Weisz PB (1980) Pure Appl Chem 52:2091

Whan DA (1981) Chemistry and Industry 532

7 Methanation

The formation of methane from $CO + H_2$ (methanation) has been used for many years to remove traces of CO (and CO_2) from hydrogen rich industrial gases such as those used for the ammonia synthesis. Renewed interest stems from the efforts to find new coal-based routes to energy sources, in this particular case to SNG (substitute natural gas). *Nickel is the preferred catalyst.*

The great scientific interest in methanation is obvious also for another reason: apparently the reaction, taking place with the simple stoichiometry

$$CO + 3H_2 \rightarrow CH_4 + H_2O \qquad\qquad (7.1)$$

promises an easy access to the large family of catalytic CO hydrogenation reactions. However, the appearance is deceitful, as it will become clear in the course of this Chapter. There are at least two major routes to CH_4, one of them presenting almost the whole manifold of reaction steps involved in the catalytic formation of higher hydrocarbons and primary monoalcohols.

7.1 Carbide Mechanism

Nickel powder reacts with carbon monoxide at $250° - 270 °C$ giving nickel carbide, Ni_3C; the carbidic carbon can be transformed quantitatively to methane and small amounts of higher hydrocarbons, by hydrogen at $250-300 °C$ (Bahr and Bahr, 1928; cf. Sect. 3.4 and 4.1).

Presumably it was this old and well based knowledge that led many investigators to the belief that the formation of metal carbide, followed by its hydrogenation to CH_x ($x = 1, 2, 3, 4$) is the main route to CH_4 (and by combination of CH_x with $x < 4$ to higher hydrocarbons), for nickel as well as for other transition metal catalysts (see e.g Biloen et al., 1979; Ekerdt and Bell, 1979; McCarty and Wise, 1979; Bonzel and Krebs, 1980; Somorjai, 1981; Brady and Pettit, 1981; Solymosi et al., 1982; Vannice and Twu, 1983).

However, there is increasing evidence that the carbide mechanism is not the only route to CH_4, and certainly not the main route to higher hydrocarbons. In this Chapter we shall discuss some of the evidence, as far as it refers to methanation; we shall pursue the discussion in context with the Fischer-Tropsch synthesis (Chap. 9).

Rabo et al. (1978), using pulse reactor techniques, have observed that carbon monoxide is almost quantitatively disproportionated to C and CO_2 on a Ni surface at 300 °C; the carbon reacts readily with a pulse of hydrogen to give CH_4, at 300 °C and even at room temperature. However, the authors detected also that cobalt at 180 °C led to a 2:1 mixture of adsorbed CO and deposited C, and hydrogen pulses produced considerably more CH_4 than deposited carbon would have accounted for. Moreover, essentially no carbon was deposited on a palladium catalyst at 300 °C, but 1/3 of the adsorbed CO molecules was transformed to CH_4 with one single hydrogen pulse.

Similar differences between various metal catalysts appeared from the work of Sachtler et al. (1979), who used evaporated films as catalysts, covered them with almost a monolayer of ^{13}C by disproportionation of ^{13}CO, and brought the surfaces in contact with $^{12}CO/H_2 = 1/5$ mixtures afterwards, at 250 °C. The formation of $^{12}CH_4$ and $^{13}CH_4$ was followed mass-spectroscopically. On the Ni-surface, the initial rate of formation of $^{13}CH_4$ (in the first 20 min) was at least an order of magnitude higher than that of $^{12}CH_4$, indicating that in this case ^{13}C rather than incoming ^{12}CO is used for the methanation. With Co, the two rates were comparable, and with Ru, the production of $^{12}CH_4$ was faster than that of $^{13}CH_4$. It was suggested that either different routes to CH_4 are involved, or carbon ^{13}C undergoes "deactivation" on Ru more than on Co and Ni, and that the supply of fresh ^{12}C by disproportionation of ^{12}CO is more important on Ru than on Ni. Since, however, the disproportionation of CO in the absence of hydrogen was found slower on Ru than on Ni and Co, the latter interpretation appears less plausible.

These data seem to indicate that the carbide route to methane may be important in the case of nickel, whereas on Co, Pd and Ru catalysts other mechanism(s) play an equally great or even the predominant role.

However, not even the nickel case is as straightforward as it may appear from the foregoing discussion. This is shown by the following investigations:

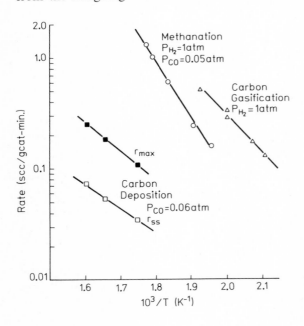

Fig. 7.1. Comparison of the methanation rate with the rates of carbon deposition and carbon gasification (= hydrogenation) for a 2% Ni/SiO₂ catalyst (Ho and Harriott, 1980). Reproduced by permission of Academic Press

To clarify the role of carbon deposition and C hydrogenation in the synthesis of CH_4 on nickel, Ho and Harriott (1980) compared the rates of these steps separately with that of the steady state methanation from $CO + H_2$, on a Ni/SiO_2 catalyst, under similar conditions ($p_{H_2} = 1$ atm and/or $p_{CO} = 0.05-0.06$ atm), as a function of the *temperature*. Figure 7.1 shows these data. The most significant result is that the rate of carbon deposition was much lower than that of CH_4 formation. The authors had shown that the rate of carbon deposition goes through a maximum (r_{max}) and comes to a steady state after a few minutes (r_{ss}) . Both rates are shown in Fig. 7.1. But even the maximum rate is too low to be of any importance. The carbon gasification rate was found higher than the rate of steady state methanation. However, it should be taken into account that this rate evidently depends on the amount of carbon available on the surface (in the particular case represented in Fig. 7.1 it was $\simeq 1/3$ of a monolayer). The authors suggested that, since the simple dissociation is too slow to be important, a dissociation of adsorbed CO with the "participation of adsorbed hydrogen" according to

$$CO + 2H \rightarrow C + H_2O \tag{7.2}$$

might be rapid enough to explain the higher methanation rate.

Mori et al. (1982), also working with a Ni catalyst, but using pulse techniques, discovered that the rates of formation of CH_4 and H_2O were identical, if a CO pulse was introduced into flowing H_2 (Fig. 7.2). If a pulse of O_2 was injected instead of CO, H_2O was immediately produced at a much higher rate, as shown also in Fig. 7.2. These data together are convincing evidence that all reactions after the opening of the CO bond are rapid. (This is in contrast to the opinion of most advocates of the carbide theory, who assume rapid carbon deposit followed by slow, rate-determining hydrogenation.) Moreover, Mori et al. found an inverse H/D isotope effect for the formation of CH_4. The mere fact of a difference of rate if H_2 or D_2 are used indicates that the simple dissociation of CO cannot be the slow and rate determining step, the presence of an inverse H/D effect excludes the H-assisted CO bond

Fig. 7.2. Amounts of CH_4 ($\square$) and H_2O ($\circ$) produced if a pulse of CO is introduced into flowing H_2 over a 20% Ni/Al_2O_3 catalyst at 200 °C. Closed circles show the amount of H_2O produced when O_2 is pulsed instead of CO (Mori et al., 1982). Reproduced by permission of the American Chemical Society

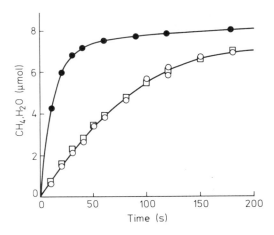

breaking mechanism (Eq. 7.2, or similar hydrogen-assisted $C-O$ breaking mechanisms, see e.g. Vannice and Twu, 1983), because a normal kinetic isotope effect would be expected for such a reaction which involves the addition of hydrogen atoms. (A more detailed discussion of inverse H/D effects will be given in Sect. 7.3). Mori et al. concluded that the CO molecule is partially hydrogenated before $C-O$ bond dissociation takes place. A hydroxycarbene ligand was considered as the key intermediate. (This suggestion had been made earlier, see Vannice, 1975, and references therein.)

The results of Mori et al. may appear to contradict those of Rabo et al. and Sachtler et al. mentioned above, since in the latter works the formation of CH_4 from carbidic carbon and hydrogen was clearly demonstrated. However, it should be born in mind that the conditions were quite different. In the experiments of Mori et al., hydrogen and a small amount of CO were admitted to a fresh Ni surface. Thus, the Mori et al. data, while showing the existence of a route to CH_4 different from the carbide route even in the Ni case, do not necessarily rule out the possibility that under more severe conditions (higher CO content, higher temperature) the carbide route to CH_4 may play a more or less important role in steady state methanation on nickel catalysts (and to a lesser degree on cobalt, but not on palladium or ruthenium).

Goddard et al. (1977) tried to estimate the energetics of the carbide route to methane, on Ni, by theoretical methods. Calculated bond energies for possible intermediates were used to obtain thermochemical information for different reaction routes. Assuming that the carbon atom resulting from the decomposition of CO is bonded to three Ni, carbon deposition would be endothermic by 67 kJ/mol, from which it was concluded that deposited carbon is not monoatomic. Only for a C_2 species bonded to four Ni in a distorted ethylenic type structure (but not for C_3 or C_4 species) did the authors find a less unfavorable reaction for adsorbed CO:

$$CO_{ad} \rightarrow 1/2\,(C_2)_{ad} + O_{ad} \qquad \Delta H = +8 \text{ kJ/mol}.$$

For this hypothetic dimer, all further hydrogenation steps are favorable:

$$
\begin{array}{c}
\text{Ni} \\
\quad \diagdown \\
\text{Ni}-\text{C} \\
\quad \| \quad + \text{H}_2 \quad \longrightarrow \\
\text{Ni}-\text{C} \\
\quad \diagup \\
\text{Ni}
\end{array}
\qquad
\begin{array}{c}
\text{Ni} \quad \text{H} \\
\quad \diagdown \diagup \\
\text{Ni}-\text{C} \\
\quad | \\
\text{Ni}-\text{C} \\
\quad \diagup \diagdown \\
\text{Ni} \quad \text{H}
\end{array}
\qquad -175 \text{ kJ/mol}
$$

$$
\begin{array}{c}
\text{Ni} \quad \text{H} \\
\quad \diagdown \diagup \\
\text{Ni}-\text{C} \\
\quad | \quad\quad + \text{H}_2 \quad \longrightarrow \\
\text{Ni}-\text{C} \\
\quad \diagup \diagdown \\
\text{Ni} \quad \text{H}
\end{array}
\qquad
\begin{array}{c}
\text{Ni} \\
\quad \diagdown \\
\text{Ni}-\text{CH}_2 \\
\quad | \\
\text{Ni}-\text{CH}_2 \\
\quad \diagup \\
\text{Ni}
\end{array}
\qquad -79 \text{ kJ/mol}
$$

$$
\begin{array}{c}
\text{Ni} \\
\quad \diagdown \\
\text{Ni}-\text{CH}_2 \quad + \text{H}_2 \quad \longrightarrow
\end{array}
\quad
\begin{array}{c}
\text{Ni}-\text{H} \\
\\
\text{Ni}-\text{CH}_3
\end{array}
\qquad -29 \text{ kJ/mol}
$$

$$\text{Ni}-\text{CH}_3 + \text{H}_2 \longrightarrow \text{Ni}-\text{H} + \text{CH}_4 \qquad -21 \text{ kJ/mol}$$

Goddard et al. suggested that if the carbide route to methanol on Ni catalyst is operative it should involve the C_2 species bonded to four Ni atoms.

The iron case is somewhat different. It is known since the old days of the Fischer-Tropsch synthesis that iron catalysts are transformed to bulk iron carbide in the course of the catalytic CO hydrogenation (see e.g. Storch et al., 1951). Raupp and Delgass (1979) measured the rate of carbide formation during methanation on 10% Fe/SiO_2 at 250 °C, by means of in situ Mössbauer spectroscopy; simultaneously the rate of CO hydrogenation (major product CH_4, some higher hydrocarbons) was determined. It was found that the rate of CO hydrogenation increases as the carburation proceeds, and reaches steady state conditions when carburation is complete. This may indicate that in the early stages part of the CO is used up for bulk carburation, and hence is not available for hydrocarbon formation, or that iron carbide is the better catalyst. Raupp and Delgass then compared the rate of steady state hydrogenation of CO ($H_2/CO = 3.3$) with the rate of formation of CH_4 when only hydrogen was led over the fully carburized catalyst. The carbide hydrogenation went through a sharp maximum near $10-12$ min and then declined quickly. Even at the highest rate observed, the rate of carbide hydrogenation was still a factor of four less than the steady state synthesis rate from $CO + H_2$. The authors concluded that either the direct hydrogenation of surface carbide is not responsible for the production of a majority of methane in the $CO + H_2$ synthesis reaction, or that a small number of very active sites are continuously replenished by CO dissociation. The latter suggestion, however, is highly improbable in view of observations by Matsumoto and Bennett (1978) who determined the rate of formation of CH_4 from $H_2/CO = 10:1$ on an iron catalyst under steady state conditions (i.e. after carburation is complete) at 250 °C. The gas flow was suddenly changed to pure hydrogen. A sharp increase of the rate of methane formations was observed, followed by a decay which was to be expected due to the shutting off of the CO supply. The sudden increase in the first few minutes is certainly incompatible with the necessity of rapid replenishing of active sites by CO disproportionation since no CO was supplied. We then are left with the first suggestion of Raupp and Delgass: that the carbide route does not play a major role in the synthesis of CH_4 from $CO + H_2$ on iron.

The rate maximum observed by Matsumoto and Bennett (1978) may merit some more attention. First of all, the phenomenon is not limited to iron, but has also been observed with Ni (Ho and Harriott, 1980). To interpret the phenomenon it is necessary to consider the steady state situation of the catalyst surface. Carbon monoxide and hydrogen occupy all available coordination sites on the surface metal atoms, whereby hydrogen and carbon monoxide, supposedly, can displace each other, according to

$$L_nM\begin{matrix} \diagup CO \\ \diagdown CO \end{matrix} \underset{CO}{\overset{H_2}{\rightleftarrows}} L_nM\begin{matrix} \diagup H \\ \diagdown H \end{matrix} \qquad (7.3)$$

as it is known in homogeneous organometallic chemistry (cf. Eqs. 4.3 and 4.4). However, the CO ligand predominates even at a relatively high excess of H_2 in the CO/H_2 gas mixture. This is born out by the empirical rate law of the CO hydrogenation

$$r \sim p_{H_2}^n \, p_{CO}^m \tag{7.4}$$

which for most conditions is operative with n near unity and m negative (cf. Sect. 5.5.2). Only at very low CO pressure ($p_{CO} = 10^{-1} - 10^{-2}$ atm), combined with high hydrogen pressure ($p_{H_2} \geqq 3$ atm), can hydrogen compete sufficiently well for the coordination sites, so that m becomes zero, and eventually even positive (see e.g. Ho and Harriott, 1980).

Returning now to the transient response experiments involving sudden change from ($CO + H_2$) to H_2 feed, we have to consider that under steady state conditions hydrogen has a rather limited access to the active metal sites due to the strong competition by CO. Nevertheless, several additions of hydrogen to carbon are necessary for the reaction to take place (cf. Eq. 7.1). A steady state rate corresponding to the prevailing conditions is observed (which is generally considered as slow compared to other heterogeneous metal catalyzed reactions, see e.g. Dautzenberg et al., 1977). If now the CO flow is interrupted, while H_2 keeps flowing, equilibrium (7.3) takes care of an increase of available hydrogen. Whatever the detailed mechanism of the transformation of CO to CH_4, the reaction rate can be expected to increase in the first moment, to decrease thereafter, as the carbon present at the catalytically active surface metal centers (as CO or in whatever intermediate form) is depleted through reaction to CH_4.

7.2 CO Insertion Mechanism

Pichler and Schulz (1970) and Henrici-Olivé and Olivé (1976) suggested reaction mechanisms for the catalytic hydrogenation of CO, the first step of which is, in both cases, the insertion of CO into a metal-hydrogen bond:

$$
\begin{array}{ccc}
H & & H \\
| & & | \\
L_n-M + CO & \rightleftarrows \; L_n-M-CO \; \rightleftarrows \; L_n-M-C{\displaystyle \mathop{\diagdown}^{\diagup O}_{H}}\,.
\end{array}
\tag{7.5}
$$

From there on the mechanisms differ. We shall discuss here the mechanism suggested by the present authors, not only because it has been adopted and corroborated in the meantime by several authors (Pruett, 1977; Feder and Rathke, 1980; Dombek, 1980; Fahey, 1981; Collman et al., 1983), but also because it is entirely based on the experiences and knowledge of coordination and organometallic chemistry (cf. Chap. 5), and because it has been supported step by step by model reactions, during the past few years.

The mechanism was suggested originally for the Fischer-Tropsch synthesis of higher hydrocarbons and primary alcohols (Henrici-Olivé and Olivé, 1976). In this Section we shall discuss only the first steps of it, as far as they relate to methanation. The description of the complete mechanism is relegated to Chap. 9, where the Pichler mechanism will also be treated for comparison.

The feasibility of Eq. (7.5), as well as evidence for its occurrence in homogeneous and heterogeneous systems, have amply been discussed in Sect. 4.2.2 and 5.3.5. In Scheme 7.1 the formation of the formyl ligand comprises steps 1a and 1b.

The next step (2) is the oxidative addition (= activation) of a hydrogen molecule to the same metal center, followed by the reductive elimination of formaldehyde (3).

Generally, no appreciable amounts of formaldehyde are found, though there is mass-spectroscopic evidence for its formation (Kölbel and Hanus, 1974; Kitzelmann et al., 1977). The most plausible explanation is that formaldehyde remains prevailingly π-bonded to the metal center. A relatively strong bond between metal center and π-bonded formaldehyde has been verified in several model complexes (cf. Sect. 4.2.3).

The next step (4) in Scheme 7.1 is the insertion of the coordinated formaldehyde into the metal$-$H bond. Here again, we are on solid grounds as far as model reactions are concerned (cf. Sect. 5.3.9).

Another H_2 molecule is then oxidatively added (step 5).

From here reductive elimination would generate methanol. In most typical methanation systems this reaction does not play any significant role; most

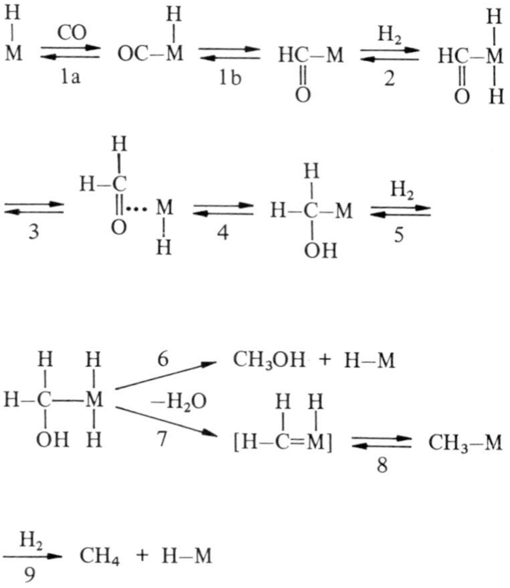

Scheme 7.1

homogeneous methanol catalysts, however, produce some methane simultaneously (cf. Chap.8). In methanation, the predominant next step is dehydration (step 7).

The intermediate formation of a carbene ligand en route to the alkyl ligand is suggested as the most plausible way to eliminate water (cf. Eq. 7.1). The suggestion is based on a large body of evidence for the existence of transition metal carbene complexes (see e.g. Fischer, 1976; Lappert and Pye, 1978; Casey, 1979), as well as their importance in catalysis (see e.g. Henrici-Olivé and Olivé, 1977). The equilibrium between carbene-transition metal hydrides and metal alkyls (step 8) is also well documented (see Sect. 5.4.2), which makes this reaction path very attractive.

A final oxidative addition of H_2 followed by reductive elimination of methane (step 9) closes the reaction cycle. Note that the metal hydride is regenerated and can start a new cycle.

Goddard et al. (1977) have checked theoretically the thermodynamic feasibility of this reaction sequence, for the case of a nickel catalyst. By calculating the bond energies of the relevant intermediates they were able to examine the energetics of the reaction cycle, with the following result:

$$Ni-H + CO \rightarrow Ni-C\overset{H}{\underset{O}{\diagdown}} \qquad \Delta H = -9 \text{ kJ/mol}$$

$$Ni-C\overset{H}{\underset{O}{\diagdown}} + H_2 \rightarrow Ni-\underset{\underset{OH}{|}}{CH_2} \qquad \Delta H = -7.1 \text{ kJ/mol}$$

$$Ni-\underset{\underset{OH}{|}}{CH_2} + H_2 \rightarrow Ni-CH_3 + H_2O \qquad \Delta H = -79 \text{ kJ/mol}$$

$$Ni-CH_3 + H_2 \rightarrow Ni-H + CH_4 \qquad \Delta H = -21 \text{ kJ/mol}.$$

The gratifying result is that the cycle is throughout favorable. Other mechanisms have been dismissed by the same method. Although this does not prove necessarily that the formyl route to methane is the only one, it certainly is a good support.

A very instructive demonstration of the stepwise reduction of coordinated carbon monoxide, via formyl and hydroxymethyl ligands to a methyl ligand, on a mononuclear model rhenium compound has been reported by Sweet and Graham (1982):

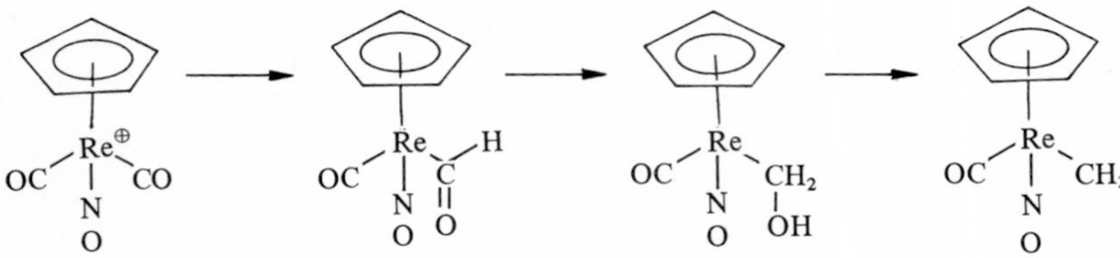

I (BF_4^- salt)   II   III   IV

The reduction was carried out with $NaBH_4$, in a mixed H_2O/THF solvent; each step was achieved with one equivalent of the reducing agent. The first two steps required 15 minutes at $0°$ each, giving $> 90\%$ yield; the last step required 5 hours at $25°C$, with 88% yield. Complexes I through IV are stable, crystalline substances that have been characterized by IR, ^1H NMR and elementary analysis. The same reaction sequence had been suggested earlier by Treichel and Shubkin (1967) for a similar molybdenum complex but the intermediates had not been detected. Reduction of ligand CO to CH_3 with $NaBH_4$ was also achieved on an iron complex, and evidence for an intermediate formyl complex was found (Lapinte and Astruc, 1983). Thorn and Tulip (1982) prepared an iridium formyl complex by oxidative addition of formaldehyde, and reduced it with $NaBH_4$ to the corresponding hydroxymethyl complex:

$$[L_4Ir]^+ \xrightarrow{CH_2O} [cis\text{-}HIr(CHO)L_4]^+ \xrightarrow{NaBH_4} [cis\text{-}HIr(CH_2OH)L_4]^+$$

($L = PMe_3$). The latter was isolated, and its crystal structure determined. The $C-O$ bond length in the hydroxymethyl ligand is significantly longer than that in ordinary alcohols. Interestingly, the further reaction to the methyl ligand (dehydration) did not require a reducing agent, but was achieved with a very weak acid, $HBF_4 \cdot$ etherate. Intermediate formation of a carbene is suggested:

$$[cis\text{-}HIr(CH_2OH)L_4]^+I^- \xrightarrow{HBF_4 \cdot ether} [L_4Ir\overset{H}{=}CH_2]^{2+}I^-BF_4^- \rightarrow [L_4Ir\overset{I}{-}CH_3]BF_4$$

All these model reactions lend plausibility to the CO insertion mechanism depicted in Scheme 7.1, which explains the *methane formation* from CO and H_2, *on a single active metal center, without violating known principles concerning the reaction patterns of transition metal centers.*

With regard to the capability of isolated mononuclear metal species to transform CO to CH_4 in a heterogeneous system, the work of Brenner and Hucul (1980) is of considerable relevance. These authors deposited many mono-, di- and polynuclear carbonyl complexes onto γ-Al_2O_3, by adsorption from pentane solutions or by sublimation, at very low concentration. During temperature programmed decomposition experiments, under H_2, they observed first desorption of part of the carbonyl ligand, followed by CH_4 formation on the remaining "subcarbonyl" species. The CH_4 evolution per metal center was essentially the same for mono-, di-, and polynuclear species. Even more important, they could exclude the sintering of mononuclear to polynuclear species by observing the CH_4 formation as a function of the catalyst loading, in the case of $Mo(CO)_6$. For a nuclearity n being required for activity, the amount of CH_4 formed, m_{CH_4}, is proportional to the number of sites of this nuclearity which would increase with the nth power of the loading w:

$$m_{CH_4} = a\ w^n$$

$$\log m_{CH_4} = \log a + n \log w .$$

A graphic evaluation of experimental data (plot log m_{CH_4} versus log w), for w varying over several orders of magnitude, resulted in a straight line, and the value of n close to unity. This is most consistent with *the activity residing in mononuclear species.*

It may be interesting to consider the coordination site requirements for such a mononuclear, active catalyst site. A look at Scheme 7.1 quickly reveals that some of the steps would require three available coordination sites. Evidently, such metal species are most probably located on steps, kinks, or surface defects. Literature evidence with other heterogeneous transition metal catalysts in fact shows that catalytic activity is confined to such special metal atoms, and that up to three free coordination sites per metal atom can be found.

Rodriguez and van Looy (1966) published electron microscopic photographs of $TiCl_3$ crystals on which propylene was polymerized. Figure 7.3 shows that the active centers, where polymer formation can be discerned are not evenly distributed over the most prominent hexagonal planes (0001) of the $TiCl_3$ crystals, but are located at the edges and along the growth spirals of the crystal which represent steps and kinks, and at surface defects.

Evidence for the availability of three coordination sites stems from CrO_3/SiO_2 catalysts. After the usual reductive treatment a considerable amount of chromium is present as Cr(II). Hogan (1970) reported that acetylene is cleanly converted to benzene over the reduced catalyst. This indicates free sites for the simultaneous coordination of three acetylene molecules around a single metal center. Garrone et al. (1975) treated the reduced catalyst with NO and studied the IR spectra of the surface species. A

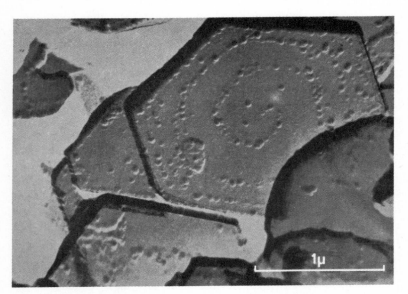

Fig. 7.3. Growing polypropylene at the edges and growth spirals of hexagonal α-$TiCl_3$ crystals (according to Rodriguez and van Looy, 1966). Reproduced by permission of John Wiley & Sons

strong doublet at 1747 and 1865 cm^{-1} was assigned to two NO molecules coordinated to the same Cr center. The spectrum was deeply affected on contacting the sample with bases such as pyridine, NH_3 or H_2O. The doublet disappeared gradually as the surface coverage with the strong base was increased. At the same time a new doublet appeared at somewhat lower wave numbers, and with comparable intensity at full coverage with the base. It was concluded that the new nitrosyl bands correspond to surface species which have taken up an extra ligand L:

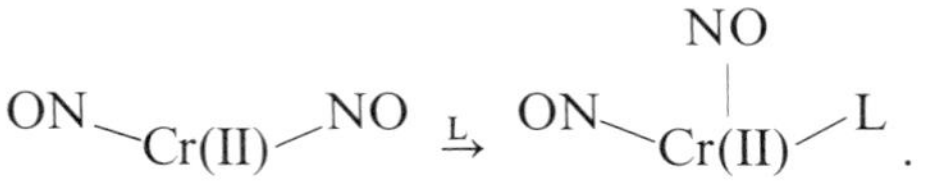

$$ON{\diagdown}_{Cr(II)}{\diagup}^{NO} \xrightarrow{L} \overset{\displaystyle NO}{ON{\diagdown}_{Cr(II)}{\diagup}L}.$$

It is, reasonably, inferred that the Cr(II) surface species possess, before adsorption, three available coordination sites.

Under the conditions of catalytic CO hydrogenation, the supposedly three available coordination sites of the active surface species are, of course, never "free", but occupied by CO and/or H_2, and reaction intermediates. Taking into account that complex formation with CO is energetically much more favorable than oxidative addition of H_2, due to the costly opening of the $H-H$ bond (431 kJ/mol):

$$M + 2CO \rightarrow M{\diagup}^{CO}_{\diagdown CO} \qquad \Delta H \gtrsim -400 \text{ kJ/mol}$$

$$M + H_2 \quad \rightarrow M{\diagup}^{H}_{\diagdown H} \qquad \Delta H \simeq -100 \text{ kJ/mol}$$

(cf. Table 3.3 and Sect. 2.2), one can visualize that the hydrogen ligands have to "fight" their entrance into the coordination sphere by means of equilibrium (7.3).

7.3 Inverse H/D Isotope Effects

The methanation on Ni catalysts is by a factor of 1.3 more rapid with D_2 than with H_2, i.e. $k_H/k_D \simeq 0.75$ (Mori et al., 1982).

Similar inverse isotope effects have been reported before: for instance, for Ni as early as 1945 (Luytens and Jungers, 1945; Nicholai et al., 1946); for Co by Sakharov and Dokukina (1961); for Ru by Kellner and Bell (1981) and Kobori et al. (1981).

Most authors interpreted this effect as the consequence of a rapid dissociation of CO followed by slow rate determining hydrogenation of C, involving a series of preequilibria. However, this assumption appears untenable in view of

Mori's observation of equal rates for the formation of CH_4 and water (cf. Fig. 7.2 and its discussion in Sect. 7.1).

In principle, the isotope effect caused by a slow, rate determining step can be estimated using transition state theory, and taking into account the zeropoint energy for all vibrations where H (or D) is involved for the starting situation and the transition state (Ozaki, 1977). Assuming that the opening of the C$-$O bond in a partially hydrogenated CO ligand is the rate determining step (Sect. 7.1), and assuming a certain geometry for the transition state, Mori et al. came up with an estimated theoretical value for k_H/k_D close to their experimental value.

The subject is, however, more complicated. Feder et al. (1981) reported an inverse isotope effect ($k_H/k_D = 0.73$) for the homogeneous high pressure hydrogenation of CO to methanol with $HCo(CO)_4$. Methanol formation takes place without opening of the C$-$O bond (see Chap. 8), and yet the same inverse isotope effect is found as in methanation. Finally, Kobori et al. (1981) reported an inverse isotope effect even for the hydrogenation of deposited carbon, on Ru.

Ozaki (1977) has pointed to the possibility that the "isotope effect on the rate of reaction does not always arise from the rate determining step but can also arise from a thermodynamic isotope effect on the concentration of an intermediate". Wilson (1979) has indicated that the adsorption coefficient for D_2 on Ni is 1.4 to 1.5 times that for H_2 at 100 to 200 °C, and that evidence for the preferential adsorption of D_2 in the presence of H_2 is also available for other metals.

In terms of coordination chemistry this means that equilibria of the type (7.3) become more favorable for D_2 than for H_2. Referring to the suggested Scheme 7.1, any of the oxidative additions of H_2 to the active center (or all of them) may be favorably affected by this phenomenon, so that the respective intermediate(s) will be present in a higher concentration in the D_2 case. The effect must be large enough to outweigh the probably normal isotope effect of the actual transfer of a hydrogen ligand to another ligand.

Wilson (1979) cautioned that it appears impossible to decide what the rate determining step in the CO hydrogenation is, simply by substituting D_2 for H_2 and noting changes in the kinetics. In other words, more discriminating experiments would be necessary. Feder et al. (1981) tried to obtain more insight by applying current methods of theoretical chemistry to the problem. Their calculations led to the conclusion that in their high pressure homogeneous cobalt system the addition of H_2 to the formyl ligand is the rate determining step. The more general statement is probably that an *inverse isotope effect is indicative of one or several preequilibrium steps in the reaction mechanism* (Collman et al., 1983).

7.4 Conclusion

In this Chapter it has been shown that for most metal catalysts the generation of methane from $CO + H_2$ is *not likely to take place predominantly via carbide*. The CO insertion mechanism, on the other hand, has a large degree of probability, as well as theoretical and experimental support. Solely in the cases of Ni and Co catalysts, the carbide route may play a certain (for nickel perhaps a major) role.

7.5 References

Bahr HA, Bahr TH (1928) Ber Dtsch Chem Ges 61:2177
Biloen P, Helle HN, Sachtler WMH (1979) J Catal 58:95
Bonzel HP, Krebs HJ (1980) Vide Couches Minces 201 (Suppl Proc Int Conf Solid Surf 4th V 1), 411
Brady RC, Pettit R (1981) J Amer Chem Soc 103:1287
Brenner A, Hucul DA (1980) J Amer Chem Soc 102:2484
Casey CP (1979) CHEMTECH, 378, and references therein
Collman JP, Brauman JI, Tustin G, Wann GS (1983) J Amer Chem Soc 105:3913
Dautzenberger FM, Helle JN, van Santen RA, Verbeek H (1977) J Catal 50:8
Dombek BD (1980) J Amer Chem Soc 102:6855
Ekerdt JG, Bell AT (1979) J Catal 58:170
Fahey DRJ (1981) J Amer Chem Soc 103:136
Feder HM, Rathke JW (1980) Ann NY Acad Sci 333:45
Feder HM, Rathke JW, Chen MC, Curtiss LA (1981) in: Catalytic Activation of Carbon Monoxide. Ford PC (ed) American Chemical Society, Washington D. C. ACS Symp No 152:19
Fischer EO (1976) Adv Organometal Chem 14:1 (Nobel Address)
Garrone E, Ghiotto G, Coluccia S, Zecchina A (1975) J Phys Chem 79:984
Goddard WA, Walch SP, Rappé AK, Upton TH, Meliu CF (1977) J Vac Sci Technol 14:416
Henrici-Olivé G, Olivé S (1976) Angew Chem 88:144; Angew Chem Int Ed Engl 15:136
Henrici-Olivé G, Olivé S (1977) Coordination and Catalysis. Verlag Chemie, Weinheim, New York
Ho SV, Harriott P (1980) J Catal 64:272
Hogan JP (1970) J Polymer Sci A-1, 8:2637
Kellner CS, Bell AT (1981) J Catal 67:175
Kitzelmann D, Vielstich W, Dittrich T (1977) Chem-Ing-Tech 49:463
Kobori Y, Naito S, Onishi T, Tamaru K (1981) J Chem Soc Chem Commun 92
Kölbel H, Hanus D (1974) Chem-Ing-Tech 46:1043
Lapinte C, Astruc D (1983) J Chem Soc Chem Commun 430
Lappert MF, Pye PL (1978) J Chem Soc Dalton 837 and references therein
Luytens L, Jungers JC (1945) Bull Soc Chim Belg 54:303
McCarty JG, Wise H (1979) J Catal 57:406
Matsumoto H, Bennett CO (1978) J Catal 53:331
Mori T, Masuda H, Imal H, Mlyamoto A, Baba S, Murikami Y (1982) J Phys Chem 86:2753
Nicholai J, Hont M, Jungers JC (1946) Bull Soc Chim Belg 55:160
Ozaki A (1977) Isotopic Studies of Heterogeneous Catalysis. Academic Press, New York
Pichler H, Schulz H (1970) Chem-Ing-Tech 42:1162
Pruett RL (1977) Ann NY Acad Sci 295:239
Rabo JA, Risch AP, Poutsma ML (1978) J Catal 53:295

Raupp GB, Delgass WN (1979) J Catal 58:361

Rodriguez LAM, van Looy HM (1966) J Polymer Sci A-1, 4:1951, 1971

Sachtler JWA, Kool JM, Ponec V (1979) J Catal 56:284 (and earlier references therein)

Sakharov MM, Dokukina ES (1961) Kinet Katal 2:710

Solymosi F, Tombácz I, Kocsis M (1982) J Catal 75:78

Somorjai GA (1981) Catal Rev-Sci Eng 23:189

Storch HH, Golumbic N, Anderson RB (1951) The Fischer-Tropsch and Related Syntheses. Wiley, New York

Sweet JR, Graham WAG (1982) J Amer Chem Soc 104:2811

Thorn DL, Tulip TH (1982) Organometallics 1:1580

Treichel PM, Shubkin RL (1967) Inorg Chem 6:1328

Vannice MA (1975) J Catal 37:462

Vannice MA, Twu CC (1983) J Catal 82:213

Wilson TP (1979) J Catal 60:167

8 Methanol from CO + H₂

The industrial production of methanol from CO and H_2 on metal oxide or metal catalysts is known since the early twenties (for a review see Falbe, 1977). Soon, it was recognized that *iron* should be eliminated from the catalyst formulations, since carburization led to desactivation of the catalyst, and to the formation of methane instead of methanol (cf. Chap. 7). The most important catalyst of the early times, and up to the late fifties, was ZnO/Cr_2O_3. The synthesis works at $250-350$ atm and $300-400\,°C$. It had been noticed early that copper containing catalysts permit the synthesis to be carried out at considerably lower pressure and temperature, but these catalysts are extremely sensitive to sulfur impurities in the feed gas. Thus, 30 ppm H_2S can be tolerated with Cr/Zn catalysts, whereas less than 1 ppm is required for Cu/Zn. The actual success of the latter catalysts is due to important improvements in the purification of the synthesis gas. Values of ≤ 0.1 ppm H_2S can be achieved today, warranting a catalyst lifetime of over three years. The Cu/Zn catalyst requires some 2% of CO_2 in the gas feed for optimum results (Klier et al., 1982). Copper (I) has been discerned as the active species (see Sect. 8.4). At CO_2 concentrations < 2% the catalyst tends to become deactivated by over-reduction, and at higher concentrations strong adsorption of CO_2 retards the synthesis. The low pressure Cu/Zn process ($50-100$ atm, $220-270\,°C$) appears to be the preferred one now, for evident reasons with regard to energy consumption. But the high pressure process is still important where sulfur (and other) impurities are a problem.

The overall reaction

$$CO + 2H_2 \overset{K}{\rightleftharpoons} CH_3OH \tag{8.1}$$

is highly exothermic (-90.8 kJ/mol). The maximum conversion is limited by the position of the equilibrium which is, of course, temperature dependent. Only the highly active Cu/Zn catalysts permit to operate the synthesis at relatively low temperatures favorable for the equilibrium (8.1).

Homogeneous catalysis of methanol formation, though possible and widely investigated, does not appear to have any industrial application in the near future, mainly because the resulting methanol is always accompanied by other products such as methyl formate and/or glycolaldehyde, as well as secondary reaction products thereof. The heterogeneous processes, on the other hand, are highly selective. Nevertheless, we shall review some of the interesting work done with the homogeneous systems in Sect. 8.2, because in the methanol case, as so often, homogeneous processes permit an easier insight into chemistry and mechanistic features of a reaction.

8.1 Nondissociative Incorporation of CO

The methanol synthesis on oxide catalysts has usually been assumed to involve nondissociated CO. But under certain reaction conditions CH_3OH is also a major product with highly dispersed supported metals as catalyst (see Sect. 8.3). In view of the controversy concerning the importance of the CO dissociation in CO hydrogenation with metal catalysts (cf. Sect. 7.1), some data of Takeuchi and Katzer (1981) are of interest. The authors provided isotopic evidence for a nondissociative reaction mechanism of methanol synthesis. A roughly 1:1 mixture of $^{13}C^{16}O$ and $^{12}C^{18}O$ was hydrogenated over a Rh/TiO_2 catalyst. At relatively low CO conversion (17.8%) the methanol formed (among other products) had an isotope distribution very close to that of the CO feed, as shown in Table 8.1.

This clearly indicates that the methanol synthesis has taken place without dissociation of CO into its atomic components. At higher conversion (27.8%) a certain amount of isotopic scrambling is observed, and in a later paper, the authors report evidence for complete isotope scrambling in the ethanol produced at 48% and 98% conversion (Takeuchi and Katzer, 1982). However, this scrambling can most probably be traced back to secondary reactions involving H_2O, due to the extremely long residence times of the reaction mixtures in a batch reactor (Henrici-Olivé and Olivé, 1984).

Another relevant experiment has been carried out by Shub et al. (1983), with a ZnO/Cr_2O_3 catalyst, and using the transient response method in a similar

Table 8.1. Isotopic compositions of methanol from CO and H_2 over Rh/TiO_2, using a 45% $^{13}C^{16}O$ – 52% $^{13}C^{18}O$ mixture (Takeuchi and Katzer, 1981).

Methanol	Composition, mol% for CO conversion	
	17.8%	27.8%
$^{12}CH_3^{16}OH$	1	12
$^{13}CH_3^{16}OH$	54	38
$^{12}CH_3^{18}OH$	44	47
$^{13}CH_3^{18}OH$	2	3

Reaction conditions: batch reactor; initial reactant pressures, $p_{CO} = 25$ torr, $p_{H_2} = 610$ torr; temperature = 150 °C.

way as described in Sect. 7.1 for the investigation of methanation. Steady state synthesis of methanol with $H_2/CO = 2$, $T = 200\,°C$ and normal pressure, was suddenly disturbed by complete suppression of the CO feed (N_2 was substituted for CO). One hour after this change the CO level in the exhaust had fallen to zero, but the formation of CH_3OH continued for over 18 hours decreasing slowly; after three hours, the rate of methanol formation was still 1/3 of that at the steady state. This observation shows that reaction intermediates are present on the catalyst surface; their initial concentration was estimated to be of the order of 1/4 of a surface layer. This finding together with kinetic data were interpreted as evidence for the presence of "CH_xO" species; the latter were assumed to form from the interaction of adsorbed CO and H. In terms of coordination chemistry, these interpretations are closely related to Scheme 7.1 in Sect. 7.2

8.2 Homogeneous Methanol Formation

Feder and Rathke (1980) found that the catalyst $HCo(CO)_4$ in benzene or dioxane solution at $\simeq 200\,°C$, with $CO:H_2 \simeq 1$, but at rather high overall pressure ($\simeq 300$ atm), permitted the homogeneous hydrogenation of CO. Primary products were methanol, methyl formate and, to a lesser extent, ethylene glycol; in secondary reactions higher primary alcohols were formed by homologation of CH_3OH (see Sect. 11.2), as well as other formates by transesterification of methyl formate. The ratio alcohol to formate was about four. It was independent of the H_2 pressure and not inversely proportional to CO pressure; this excluded the equilibrium

$$ROH + CO \rightleftarrows RCO_2H.$$

Addition of formaldehyde to the feed gas led to a similar product distribution. From these findings it was concluded that formaldehyde is an intermediate to all three major primary products. The authors suggested Scheme 8.1 for the formation of the primary products, and noted that it is remarkably similar to the one suggested earlier for the Fischer-Tropsch synthesis (Henrici-Olivé and Olivé, 1976; see Schemes 7.1 and 9.1), featuring the formyl ligand as well as the π-bonded formaldehyde ligand. An important difference is the ramification of the reaction sequence at intermediate *2*, to give the hydroxymethyl complex *3* or the methoxy complex *4*. The two insertion modes of formaldehyde have been discussed in Sect. 5.3.9. In that context it was mentioned that at $0\,°C$ and 1 atm the catalyst $HCo(CO)_4$ inserts formaldehyde according to the mode that leads to the hydroxymethyl complex. We see now that under more severe conditions the other mode becomes also available for the same catalyst.

The simultaneous formation of intermediates *3* and *4* (Scheme 8.1) merits a few reflections. Does this necessarily mean that the hydride ligand can migrate to either side of the carbonyl group of the formaldehyde? There is another possibility. The two radicals $\overset{*}{C}(OH)H_2$ and $\overset{*}{O}CH_3$ are almost isoenergetic, with the former about 20 kJ/mol more stable, and they isomerize intramolecularly with a relatively low activation energy of $\simeq 150$ kJ/mol

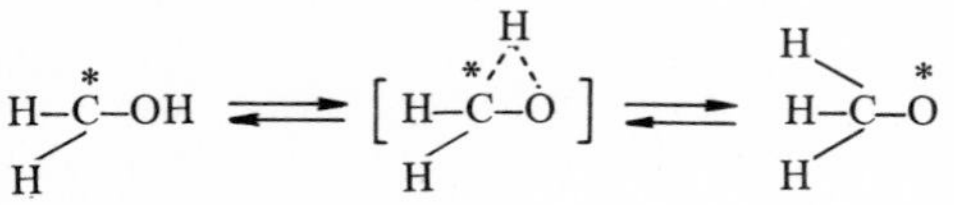

$$
\underset{H}{\overset{*}{\underset{|}{H-C}}}-OH \;\rightleftharpoons\; \left[\underset{H}{\overset{*}{H-C}}\overset{\wedge}{\underset{}{-O}} \right] \;\rightleftharpoons\; \underset{H}{\overset{H}{H-C}}\overset{*}{-O} \tag{8.2}
$$

(Saebø et al., 1983). Could this isomerization take place within the confines of a solvent cage, at elevated temperatures (but not on a heterogeneous catalyst)? It would be one possible explanation for the manifold products found in some homogeneous CO hydrogenation systems.

Almost simultaneously with Feder and Rathke, Fahey (1981) came up with essentially the same reaction mechanism (Scheme 8.1) for the same system. The intermediacy of formaldehyde was supported additionally by chemical trapping as its ethylene glycol acetal. Although the thermodynamics for the formation of formaldehyde were calculated to be unfavorable, Fahey argues that the concentration of formaldehyde permitted by thermodynamics is more than sufficient for a transient intermediate. It should also be taken into account

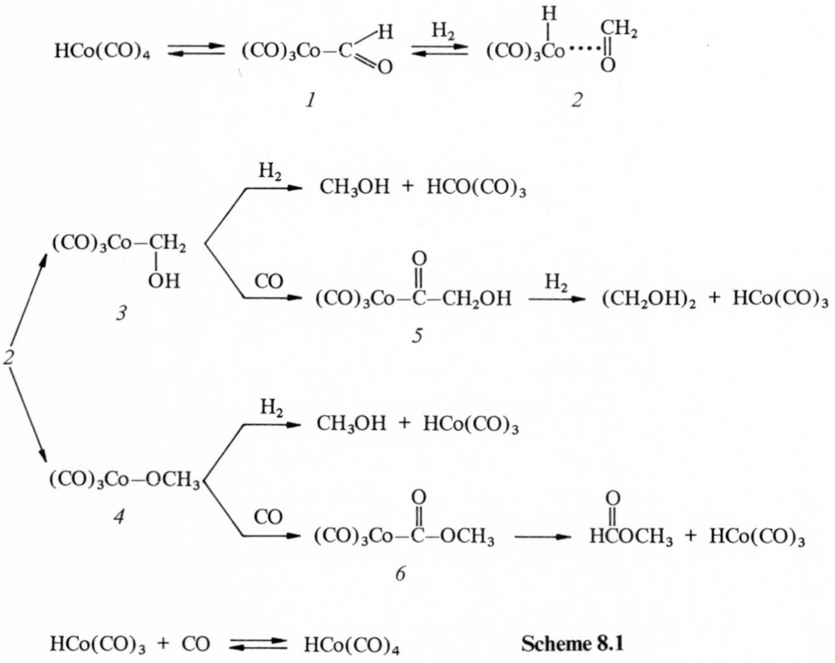

$$
HCo(CO)_4 \;\rightleftharpoons\; \underset{1}{(CO)_3Co-C\overset{\diagup H}{\underset{\diagdown O}{}}} \;\overset{H_2}{\rightleftharpoons}\; \underset{2}{(CO)_3Co\overset{H}{\underset{|}{\cdots}}\overset{CH_2}{\underset{O}{||}}}
$$

$$
2 \begin{cases}
\underset{3}{(CO)_3Co-CH_2 \atop OH} \begin{cases} \xrightarrow{H_2} CH_3OH + HCO(CO)_3 \\[6pt] \xrightarrow{CO} \underset{5}{(CO)_3Co-\overset{O}{\overset{||}{C}}-CH_2OH} \xrightarrow{H_2} (CH_2OH)_2 + HCo(CO)_3 \end{cases} \\[24pt]
\underset{4}{(CO)_3Co-OCH_3} \begin{cases} \xrightarrow{H_2} CH_3OH + HCo(CO)_3 \\[6pt] \xrightarrow{CO} \underset{6}{(CO)_3Co-\overset{O}{\overset{||}{C}}-OCH_3} \longrightarrow H\overset{O}{\overset{||}{C}}OCH_3 + HCo(CO)_3 \end{cases}
\end{cases}
$$

$$
HCo(CO)_3 + CO \;\rightleftharpoons\; HCo(CO)_4 \qquad\qquad \textbf{Scheme 8.1}
$$

that the calculations have considered free CH_2O, whereas the reaction Schemes 7.1 and 8.1 involve CH_2O that never leaves the complex, and has a relatively strong metal-carbonyl π-bond (cf. Sect. 4.2.3).

Essentially the same reaction Scheme was also adopted by Dombek (1980) for the formation of methanol on a soluble, mononuclear ruthenium carbonyl complex, at 340 atm in acetic acid. Small amounts of ethylene glycol were also found, but no formate was observed. It was assumed that with this catalyst the methoxy rather than the hydroxymethyl ligand is the precursor for methanol. It should, however, be mentioned that a ruthenium catalyst in THF solution at 1300 atm, did produce small amounts of methyl formate besides methanol (Bradley, 1979); presumably the same catalytically active mononuclear hydridocarbonyl complex was present. The intermediacy of formaldehyde was also proposed as the most plausible explanation in a system producing mainly methanol and ethylene glycol on a soluble rhodium catalyst. Parker et al. (1982) investigated the fate of ^{14}C labeled formaldehyde added to the CO/H_2 mixture in a batch experiment. The labeled formaldehyde was fully converted into a mixture of products typical for that obtained during the rhodium catalyzed conversion of CO/H_2. (The homogeneous rhodium catalysts, which are particularly useful for the synthesis of ethylene glycol and higher polyalcohols, will be further discussed in Chap. 10.)

Occasionally, small amounts of hydrocarbons have been detected in the homogeneous cobalt system (Feder and Rathke, 1980), as well as in ruthenium systems (Bradley, 1979, and references therein); however, in all cases they have been traced back to heterogeneous catalysis by precipitated metal. There remains the enigmatic question why do heterogeneous cobalt and ruthenium catalysts form hydrocarbons and soluble do not, but instead give rise to methanol, formates and glycols. We shall come back to this interesting question in Chap. 12, when the discussion of the hydrocarbon growth in Chap. 9, as well as of the particularities of hydroformylation in Chap. 10 will have provided additional insight into the mechanistic connections between all these reactions.

8.3 Methanol Synthesis with Supported Noble Metal Catalysts

Although the noble metal catalysts probably will never compete with the robust Cr_2O_3/ZnO or the highly active CuO/ZnO catalyst of the present large volume commercial methanol processes, they are very interesting from a mechanistic point of view.

Ichikawa (1978) reported a striking influence of the support on the selectivity to methanol with rhodium catalysts (Table 8.2). Working at 220 °C and atmospheric pressure ($H_2/CO \simeq 3$), the author found high selectivity to methanol with strongly basic carriers such as CaO or MgO, and almost exclusive conversion to methane with SiO_2 and Al_2O_3. These results appear to indicate a strong ligand effect exerted by the support (cf. Sect. 5.5.2 and 6.2.2).

Table 8.2. Methanol selectivity of supported Rh catalysts, according to Ichikawa (1978).

Support	T °C	CO conversion (% in 10 h)	Methanol selectivity (%)
CaO	232	8.0	98.6
ZnO	220	17.4	95.4
MgO	220	57.4	89.0
BeO	225	15.4	68.3
SiO$_2$	220	23.0	0.6
Al$_2$O$_3$	220	54.6	0.2

A possible interpretation is the assumption of an influence on the polarity of the metal-hydrogen bond, which then would lead to incorporation of formaldehyde intermediate in either of the two possible ways, as indicated in Chap. 5 (Eqs. 5.31 and 5.32):

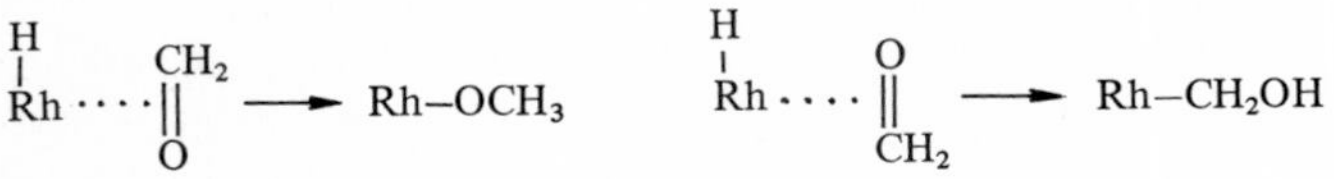

In terms of Scheme 7.1 (methanation) this would mean a ramification at step 4:

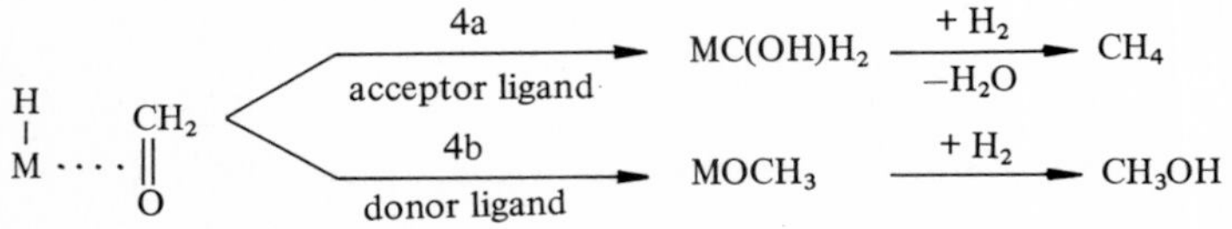

$$(8.3)$$

It may be noted that Ichikawa also found supports (La$_2$O$_3$, TiO$_2$) which, under his reaction conditions, operated simultaneously either way, giving additionally some chain growth, resulting in ethanol and higher hydrocarbons. The mechanistic features involved in chain growth will be discussed in the next Chapter.

Data of Dirkse et al. (1982) invite to a similar interpretation. The authors prepared a supported Rh catalyst with a silica which had been pretreated with Li-naphthalide to eliminate all acidity. This support was compared with untreated silica, with regard to CO hydrogenation at 40–95 atm and 225–300 °C. The Rh catalyst on the treated support gave 90–100% methanol, whereas Rh on untreated silica resulted in predominantly methane. It appears that, again, an acidic support leads to step 4a (Eq. 8.3), whereas the non-acidic support permits step 4b to take place. (Dirkse et al. suggested either a different interaction cluster-support, or a Cannizzaro reaction catalyzed by basic sites, and converting formaldehyde into methanol and formate anion.)

Amazingly, Poutsma et al. (1978) found that palladium on (untreated) SiO_2 gave methanol with high selectivity at $260-350\,°C$ and $10-1000$ atm, whereas the same catalyst produces essentially methane at normal pressure (Vannice, 1975). Poutsma et al. pointed out that high selectivity was observed only within the temperature/pressure regime for which methanol formation is thermodynamically favorable, because supported Pd also catalyzes the decomposition of CH_3OH to CO and H_2. Since the acidic SiO_2 support would suggest step 4a rather than 4b (Eq. 8.3), the selectivity to methanol should be caused by a pressure dependence of the second possible ramification of Scheme 7.1 (steps 6 and 7). For any reason, high pressure appears to prevent dehydration (step 7) in favor of reductive elimination of methanol.

8.4 Synergism in the Cu/ZnO Catalyst

The industrial low pressure $(20-100$ atm) methanol synthesis uses catalysts based on the compositions $Cu/ZnO/Cr_2O_3$ or $Cu/ZnO/Al_2O_3$. Whereas Cr_2O_3 or Al_2O_3 are considered non essential, the presence of both copper and zinc oxide is vital for satisfactory catalysis, the combination of both being several orders of magnitude more effective than either of the two components. This mutual promotion has intrigued several authors.

Herman et al. (1979) studied a wide variety of catalyst compositions, with the following results and conclusions: While after calcination the mixed catalysts consist of hexagonal ZnO and tetragonal CuO, the actual active catalyst, after reduction or after use, is made up essentially of metallic Cu and ZnO. However, there is spectral evidence for the presence of Cu(I) species which are visualized as "dissolved" in ZnO. From the known facts that Cu^+ ions form carbonyls (e.g Pasquali et al., 1981, and references therein) and that ZnO activates molecular hydrogen by heterolytic splitting, giving rise to ZnH and OH groups (Eischens et al., 1962), Herman et al. concluded that CO is coordinatively bonded to a Cu(I) center and that H_2 is heterolytically split by a nearby ZnO center, the proton attacking the carbon end and the hydride ion attacking the oxygen end of the CO molecule. (Since the polarity of CO is such that the oxygen is negative, cf. Sect. 3.1, the contrary would have appeared more probable.) Herman et al. also assumed that the synthesis proceeds as a surface reaction, because no special pore distribution was found necessary for a highly effective catalysis.

The presence of Cu(I) has been confirmed by other authors (Okamoto et al., 1982). Nevertheless, the Herman mechanism of the synergistic effect is not quite cogent. It should be taken into consideration that Cu(I) itself is well able to decompose hydrogen (see e.g. James, 1973) and also to form hydrides (Beguin et al., 1981). Moreover, Goeden and Caulton (1981) have shown that Cu(I) hydrides are more stable than Zn(II) hydrides. The latter authors have also reported several relevant observations. They investigated the catalytic reaction of a Cu(I) hydride with formaldehyde and found

$$2\,H_2CO \xrightarrow{\text{(HCuP)}_6} \overset{\displaystyle O}{\overset{\|}{HC}}-OCH_3 \tag{8.4}$$

(P = P(tol)$_3$), which indicates the intermediate formation of copper methoxide:

$$1/6\,(HCuP)_6 + H_2CO \rightarrow 1/n\,(PCuOCH_3)_n. \tag{8.5}$$

This was confirmed by using (DCuP)$_6$, and finding all of the deuterium in the methyl group:

$$H_2CO \xrightarrow{\text{(DCuP)}_6} \underset{\text{(stoichiometric)}}{HCO_2CH_2D} + \underset{\text{(catalytic)}}{HCO_2CH_3} \tag{8.6}$$

Goeden and Caulton found also that a copper alkoxide reacted with H$_2$ to give copper hydride and the alcohol:

$$1/4\,(CuOR)_4 + P + H_2 \rightarrow 1/6\,(HCuP)_6 + ROH \tag{8.7}$$

(R = t-Bu). This formal heterolytic splitting of hydrogen:

$$Cu^+ + H_2 \rightarrow CuH + H^+ \tag{8.8}$$

was considered a unique characteristic of copper.

Based on these findings, the present authors have suggested a mechanism for the amazing synergism of Cu(I) and ZnO in the methanol synthesis (Henrici-Olivé and Olivé, 1982), which is somewhat different from that proposed by Herman et al. (1979). It was suggested that the most important effect of ZnO is to stabilize the catalytically active Cu(I) state, for instance in the form of "end groups" of ZnO chains which may formally be written as

$$Cu(I)-O-Zn(II)-O-Zn(II)-O- \tag{8.9}$$

(Actually, the structure is, of course, tridimensional.) At the surface such Cu(I) species could easily coordinatively bind one or several CO molecules, under the reaction conditions of the methanol synthesis.

Translating the findings given in Eqs. (8.7) and (8.8) to the Cu(I)−O bond in Eq. (8.9), the following formation of a surface Cu(I) carbonyl hydride appears amenable:

$$(CO)_xCu-O-Zn \cdots + H_2 \rightarrow (CO)_x\overset{\overset{\displaystyle H}{|}}{Cu} + HO-Zn \cdots . \tag{8.10}$$

Equation (8.10) takes also care of Goeden and Caulton's finding that Cu(I) hydrides are more stable than Zn(II) hydrides.

The next step would be ligand CO insertion into the Cu−H bond with the formation of a formyl ligand, as it was assumed in the Schemes 7.1 and 8.1.

$$(CO)_x\overset{\overset{\displaystyle H}{|}}{Cu} + HO-Zn \cdots \rightarrow (CO)_{x-1}\overset{\overset{\displaystyle CO}{|}}{\underset{}{\overset{\displaystyle H}{|}}}{Cu} + HO-Zn \cdots . \tag{8.11}$$

Formyl ligands have been reported to transform to formaldehyde or methanol under the influence of proton donors (Collman and Winter, 1973; Casey and Neumann, 1978; Gladysz and Tam, 1978). The amphoteric character of zinc hydroxide invites the assumption that proton donation to the coordinated formyl ligand takes place, given the ideal proximity of the surface HO−Zn group:

$$\overset{\overset{\displaystyle H}{|}}{\underset{}{\overset{\displaystyle CO}{|}}}{Cu} + HO-Zn \cdots \quad \overset{CH_2}{\underset{O}{\|}} \cdots Cu-O-Zn \cdots . \tag{8.12}$$

The reaction sequence suggested in Eqs. (8.10)−(8.12), and involving the restoration of the original Cu−O−Zn arrangement, has a precedent in a soluble palladium hydrogenation catalyst involving the quadridentate ligand SALEN (Henrici-Olivé and Olivé, 1975/1976):

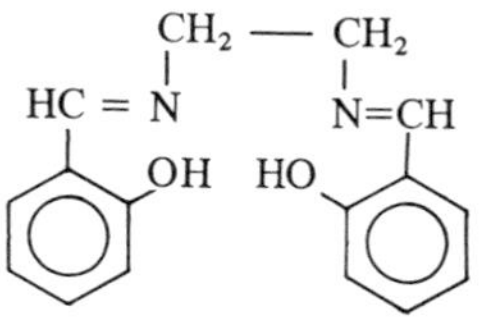

SALEN

In that case, kinetic data and the pH dependence of the reaction had strongly suggested a heterolytic splitting of the hydrogen molecule, with the proton being taken up by one of the phenoxy ligands of the complex and the H$^-$ by the metal, with a polar transition state:

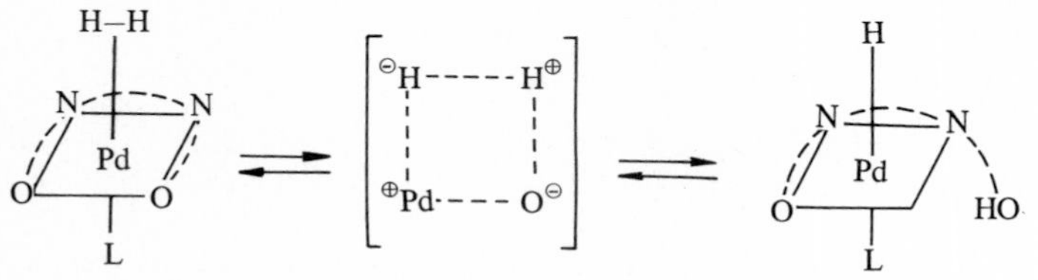

After the coordination of a substrate (olefin) molecule at the free coordination site, hydrogenation takes place. The required steric proximity of the OH group is, in this case, guaranteed by the fact that the SALEN ligand remains fixed to the metal center by three of its four coordinating groups. The product molecule then leaves the complex, and the starting configuration is restored. It should also be recalled that, in Nature, the action of hydrogenases appears to follow similar reaction patterns.

The further hydrogenation of the coordinated formaldehyde may be visualized along the same lines, involving heterolytic splitting of H$_2$ and formation of a methoxy ligand, which then is hydrogenated to the product methanol:

$$(8.13)$$

The chosen insertion mode of CH$_2$O into the Cu$-$H bond (methoxy ligand rather than hydroxymethyl ligand) follows the findings of Goeden and Caulton (Eqs. 8.4$-$8.6).

It has been mentioned earlier in this Chapter that the Cu/ZnO catalysts are very sensitive towards poisoning by H$_2$S or HCl. The poisoning may be expected to take place according to:

$$Cu-O-Zn \cdots + H_2S \rightleftarrows CuSH + HO-Zn \cdots$$

$$Cu-O-Zn \cdots + HCl \rightleftarrows CuCl + HO-Zn \cdots .$$

It appears, then that the synergistic effect of ZnO, rather than providing the activation of H$_2$ directly, has the dual purpose of maintaining the copper in the active Cu(I) state, and of providing a polar $Cu-O$ bond available for the heterolytic splitting of the hydrogen molecule.

8.5 Conclusions

The remarkable *selectivity* of the heterogeneous commercial methanol processes appears to rest upon the selective insertion of the formaldehyde intermediate into a metal-hydrogen bond according to the methoxy-ligand-forming mode (e.g. Eqs. 8.5 and 8.13; for the Cr_2O_3/ZnO catalysts this question appears not to have been investigated yet). The insertion of CO into the bond between the metal and the oxygen of the methoxy ligand has not been observed ever in heterogeneous systems (though it does take place in solution, see Scheme 8.1). Thus, *reductive elimination of the methoxy ligand together with a hydrogen ligand* is the only possible reaction.

The heterogeneous noble metal catalysts, as well as the soluble systems, on the other hand, have generally the whole variety of options discussed in this Chapter, leading to methane, methanol, formates, glycols, even sometimes higher hydrocarbons and alkanols. While the product composition can be influenced by the metal itself, the support if applicable, pressure, temperature, H_2/CO ratio and other variables, only very seldom can high selectivity to one product be achieved.

8.6 References

Beguin B, Denise B, Sneeden RPA (1981) J Organometal Chem 208:C 18
Bradley JS (1979) J Amer Chem Soc 101:7419
Casey CP, Neumann SM (1978) J Amer Chem Soc 100:2544
Collman JP, Winter SR (1973) J Amer Chem Soc 95:4089
Dirkse HA, Lednor PW, Versloot PC (1982) J Chem Soc Chem Commun 814
Dombek BD (1980) J Amer Chem Soc 102:6855
Eischens RP, Pliskin WA, Low MJD (1962) J Catal 1:180
Fahey DR (1981) J Amer Chem Soc 103:136
Falbe J (ed) (1977) Chemierohstoffe aus Kohle. Georg Thieme Verlag, Stuttgart
Feder HM, Rathke JW (1980) Ann NY Acad Sci 333:45
Gladysz JA, Tam W (1978) J Amer Chem Soc 100:2545
Goeden GV, Caulton KG (1981) J Amer Chem Soc 103:7354
Henrici-Olivé G, Olivé S (1975/76) J Molecular Catal 1:121
Henrici-Olivé G, Olivé S (1976) Angew Chem 88:144; Angew Chem Int Ed Engl 15:136
Henrici-Olivé G, Olivé S (1982) J Molecular Catal 17:89
Henrici-Olivé G, Olivé S (1984) J Phys Chem (in press)
Herman RG, Klier K, Simmons GW, Finn BP, Bulko JB, Kobylinski (1979) J Catal 56:407
Ichikawa M (1978) Bull Chem Soc Jpn 51:2268, 2273
James BR (1973) Homogeneous Hydrogenation. Wiley, New York
Klier K, Chatikavanij V, Herman RG, Simmons GW (1982) J Catal 74:343

Okamoto Y, Fukino K, Imanaka T, Teranishi S (1982) J Chem Soc Chem Commun 1405
Parker DG, Pearce R, Prest DW (1982) J Chem Soc Chem Commun 1193
Pasquali M, Floriani C, Gaetani-Manfredotti A, Guastini C (1981) J Amer Chem Soc
    103:185, and references therein
Poutsma ML, Elek LF, Ibarbia PA, Risch AP, Rabo JA (1978) J Catal 52:157
Saebø S, Radom L, Schaefer HF (1983) J Chem Phys 78:845
Shub FS, Kuznetsov VD, Belysheva TV, Temkin MI (1983) Kinet Katal 24:385
Takeuchi A, Katzer JR (1981) J Phys Chem 85:937
Takeuchi A, Katzer JR (1982) J Phys Chem 86:2438
Vannice MA (1975) J Catal 37:449, 462

9 Fischer-Tropsch Synthesis

The Fischer-Tropsch synthesis is essentially a *polymerization reaction* — or perhaps better an *oligomerization*, since in most cases the average molecular weight of the product is not very high — where carbon-carbon bonds are formed between C atoms proceeding from carbon monoxide, under the influence of hydrogen and a metal catalyst, and with elimination of water. Without anticipating the detailed reaction mechanism, the main reaction of the Fischer-Tropsch (FT) synthesis may be formulated as:

$$n(CO + 2H_2) \xrightarrow{\text{Cat}} -(CH_2)_n- + nH_2O. \tag{9.1}$$

Depending on catalyst and reaction conditions, the products are linear hydrocarbons, oxygenated derivatives thereof, or mixtures of both. Usually, a wide range of molecules with different chain length, and with a molecular weight distribution characteristic for polymerization reactions, are formed.

In compliance with the purpose and scope of this book, we shall concentrate mainly on *chemistry and mechanism* of the FT synthesis. The major part of the Chapter will be dedicated to the application of principles of polymer chemistry, as well as concepts originating in the field of molecular catalysis with defined transition metal compounds, to the experimental facts of the FT synthesis. The evident aim is the discussion of a reaction mechanism which is in accord with the mentioned principles and concepts.

However, the story of the FT synthesis would be incomplete without at least a brief account of its historical development, and of the present status of technical progress.

9.1 Early Developments and Present State of the Commercial Fischer-Tropsch Synthesis

Great discoveries in chemistry are often the consequence of accidental observations. Thus, the invention of the Ziegler polymerization of ethylene at low pressure was triggered, when residual nickel catalyst in an autoclave led to an unexpected dimerization of ethylene to 1-butene, whereas actually higher olefins were expected on an aluminum alkyl catalyst. In search of an elucidation of the "nickel effect", other metal compounds were tested, until the classical Ziegler catalyst, $TiCl_4/AlR_3$, emerged (Ziegler et al., 1955). The OXO reaction (hydroformylation) was discovered by Roelen, while searching for the synthesis of hydrocarbons in homogeneous medium (Roelen, 1948).

But not so with the FT synthesis. When the "Kaiser Wilhelm Institut für Kohleforschung" in Mülheim, Germany, was founded in 1914, the general idea of producing oil from coal was already taken into consideration. But World War I delayed this long-range research goal.

O. Roelen, who was one of the coworkers of F. Fischer and H. Tropsch, has described the development of these early days (Roelen, 1978). Since 1913, the direct liquefaction of coal according to Bergius was known. Also in 1913, the BASF had obtained patents for the production of hydrocarbons and, mainly, oxygenated derivatives thereof, from carbon monoxide and hydrogen under high pressure, with alkali activated cobalt and osmium catalysts. However, this work was not pursued by the BASF. Nevertheless, F. Fischer, then director of the Institute, decided in 1919 that he and his Institute would investigate the indirect liquefaction of coal, via synthesis gas. Repeating and developing the BASF patents, Fischer and Tropsch (1923, 1924) disclosed the production of "Synthol", a mixture of oxygen containing derivatives of hydrocarbons, produced from CO and H_2 at > 100 atm and 400 °C, with alkali treated iron shavings as catalyst. However, the catalyst was deactivated quickly by carbon deposition. Moreover, the aim was the production of hydrocarbons (as a substitute for natural gasoline) and not oxygenates. Making intelligent use of a report by Patard (1924), who had found the synthesis of methanol from CO and H_2 on a long-lived ZnO catalyst, Fischer, Tropsch and their coworkers synthesized in 1925 for the first time small amounts of ethane and higher hydrocarbons under normal pressure, at 370 °C, on a Fe_2O_3/ZnO catalyst. *This was the birth hour of the Fischer-Tropsch synthesis.* The following years saw the laborious development of improved catalysts until, in, 1934, the FT synthesis was ready for transfer to Industry. The Ruhrchemie AG was the first to be licensed to build a commercial plant, working at normal pressure and with a cobalt/thorium oxide catalyst.

The further development has been described in detail by Pichler, another coworker of Fischer and Tropsch at that time (Pichler and Krüger, 1973). Only some highlights here: Fischer and Pichler (1939) reported the synthesis at medium pressure (5–30 atm), with cobalt catalysts, which resulted in higher yields, higher molecular weight product and longer life of the catalyst. During World War II, nine Fischer-Tropsch plants were operating in Germany, based on the medium pressure cobalt catalyzed process. From 1937 on, the research shifted back to iron catalysts. Fischer and Pichler (1937/51) had discovered that alkalized iron catalysts, at medium pressure, resulted in further improvements of product yields and catalyst lifetime; moreover, the iron catalysts permitted a broader variation of the process variables and hence of reaction products. However, these important discoveries did not mature to technical utilization before the end of the war.

After 1945, there was general access to the German research and development work through the various FIAT, BIOS reports, and other publications. The Bureau of Mines, in USA, obtained, together with plans and documents, an entire Fischer-Tropsch pilot plant, and in the late forties several test facilities were built by the Bureau of Mines as well as by the private industry. (For reviews see Storch et al., 1951; Lee, 1982.)

But then, in the early fifties, the world price of the oil began to fall, and the great euphoria of cheap energy from oil exploded, not only in USA but also in

Europe. Coal mines were shut down. Apart from price, the ease of automated transport systems and feed lines for refineries, production plants, industrial and domestic heating, etc., helped oil to oust the dirty and inconvenient raw material coal.

Until the big shake-up came in the early seventies, with the oil embargo, and the escalation of the crude oil prices in its aftermath. Scientific and industrial investigation reverted to coal as a source of carbon, and to the Fischer-Tropsch synthesis as one of the most potent ways to transform it into gasoline, oil and other useful chemicals. Although, even with todays crude oil prices, the coal based routes appear to be not yet quite competitive economically, a sound effect of the embargo shock has been to create a feeling of awareness and responsibility for the future. The eventual exhaustion of petroleum sources has been predicted for a considerable period of time. Commercial processes are by now available to get a coal-to-fuel industry started, although advanced technologies, improving efficiency and reducing costs, need to be developed.

The technically important developments after 1945 are based on the work of Fischer and Pichler (1937/51) concerning the FT synthesis at medium pressure, with alkalized iron catalysts.

The only large commercial facility for the production of gasoline, gas oil, and paraffins from coal, by the Fischer-Tropsch process, is located in Sasolburg, in South Africa. This country disposes of large coal resources, but has no oil. In 1955, the South African Coal, Oil and Gas Corporation started the production of synthesis gas $(CO + H_2)$ from coal using a Lurgi coal gasification process, and the synthesis of hydrocarbons by the Fischer-Tropsch reaction (Dry and Hoogendoorn, 1981). Two parallel process designs for the FT synthesis are used in this plant (Sasol I), the fixed-bed reactor and the circulating fluid-bed reactor. The fixed-bed reactor uses a precipitated, extruded iron catalyst. The working pressure is ca. 25 atm at $220-240\,°C$; the synthesis gas is fed at a ratio $H_2/CO = 1.8$. The fluid-bed reactor works at a higher temperature $(310-340\,°C)$, at similar pressure (24 atm), and at a H_2/CO ratio of 6. The low particle size iron catalyst is moved together with the synthesis gas through the reactor in an upward gas stream and, after separation of the reaction products, is recycled. Tables 9.1 and 9.2 compare the

Table 9.1. Comparison of the products obtained with fixed-bed catalyst and with circulating catalyst (in weight %), (Pichler and Krüger, 1973).

Fraction	Fixed-bed catalyst $(220-240\,°C)$	Circulating catalyst $(310-340\,°C)$
C_3-C_4	5.6	7.7
C_5-C_{11} (gasoline)	33.4	72.3
Gas oil	16.6	3.4
Paraffin m.p. $< 60\,°C$	22.1	3.0
Paraffin m.p. $95-97\,°C$	18.0	–
Alcohols, ketones	4.3	12.6
Acids	traces	1.0

Table 9.2. Comparison of the composition of the liquid fraction for fixed-bed catalyst and circulating catalyst (in vol.%), (Pichler and Krüger, 1973).

Components	Fixed-bed catalyst		Circulating catalyst	
	C_5-C_{10}	$C_{11}-C_{18}$	C_5-C_{10}	$C_{11}-C_{18}$
Olefins	50	40	70	60
Paraffins	45	55	13	15
Oxygen containing Compounds	5	5	12	10
Aromatics	–	–	5	15

reaction products obtained by the two processes. The fixed-bed process leads to a higher average molecular weight of the product, and to 95% linear molecules; the fluid-bed process results in a higher olefin content, but has 40% of branched material (mainly methyl branches). Oxygen containing compounds and aromatics are insignificant in the former, but not in the latter process. The straight-chain material of the fixed-bed process is a useful feedstock for the production of chemicals such as plasticizers, detergents with high biodegradability, synthetic lubricants, etc., whereas the fluid-bed process is more useful for the production of gasoline. In fact Sasol I operates an intimate combination of both processes, permitting a flexible adaptation to market requirements.

The overall yield of gasoline and/or diesel can be improved by a further workup of the primary FT products, using known technology such as oligomerization of light olefins, hydrocracking, reforming, etc.

Sasol I has provided the South Africans with a unique experience which has encouraged them to build two much larger facilities for the FT process (Sasol II and Sasol III) which, by 1984, are expected to provide half of that country's liquid fuel needs. Both of these two plants will operate according to the fluid-bed process (Lee, 1982).

Aside from these large industrial plants, there appears to be scientific and industrial research going on at many places. Although the FT synthesis is but one of the various processes for the transformation of coal to a useful feedstock, it will certainly follow attracting much interest in the future.

9.2 The Products of the FT Synthesis

9.2.1 Primary Products

The product mixture contains mainly olefins and paraffins, but also variable amounts of alcohols, aldehydes, acids, esters and aromatic compounds, according to the reaction conditions. In the early days of the process it was already found that the main products obtained with cobalt catalysts were linear olefins and paraffins, and that the small amount of nonlinear products consisted predominantly of monomethyl branched compounds. Since olefins and paraffins were found to have the same skeleton, it was assumed that the paraffins are formed from olefins by subsequent hydrogenation (Tropsch and Koch, 1929; Koch and Hiberath, 1941).

The olefin component of the hydrocarbons consists mainly of α- and β-olefins (1- and 2-alkenes). The fact that the concentration of α-olefins is generally much higher than that corresponding to the thermodynamic equilibrium led Friedel and Anderson (1950) to the conclusion that α-olefins should in fact be the *primary products* of the synthesis.

This conclusion was unequivocally confirmed by Pichler et al. (1967). These authors investigated the product composition as a function of the residence time of the synthesis gas on the catalyst, varying the space velocity. The synthesis was carried out at normal pressure and 200 °C, with a precipitated cobalt/thorium oxide/kieselguhr catalyst, at a H_2/CO ratio of 2. Under these conditions the main product consists of hydrocarbons in the gasoline and gas oil range. The result of the gas chromatographic analysis of the products is

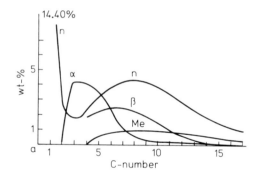

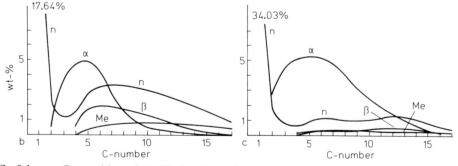

Fig. 9.1 a–c. Composition of the Fischer-Tropsch mixture of hydrocarbons as a function of the space velocity: **a)** 78 h^{-1}, **b)** 337 h^{-1}, **c)** 2380 h^{-1}. n = n-paraffins; α = α-olefins; β = β-olefins; Me = methyl paraffins. Catalyst = Co/ThO$_2$; after Pichler et al. (1967) (The space velocity is defined as the volume of synthesis gas per volume catalyst and hour)

represented in the Fig. 9.1. The clear-cut answer: the shorter the residence time, the larger is the α-olefin fraction of the product. At longer residence time (low space velocity) the primarily formed α-olefins are transformed in secondary reactions into β-olefins, linear paraffins and methyl branched products. Analogous results were also reported for the synthesis with iron catalysts at medium pressure and 220 °C (Pichler and Schulz, 1970). The mechanistic implications of these findings cannot be overestimated. Whatever reaction mechanism might be suggested, it has to take into account that the main primary reaction product consists of linear molecules with one, and only one, double bond in α position. Under the prevailing reaction conditions the paraffinic products were formed by subsequent hydrogenation of the olefins, and not by hydrogenolysis of a metal-carbon bond.

In contrast to the primary and secondary products mentioned, a small alcohol fraction found under the given reaction conditions was independent of the space velocity. Pichler and his coworkers concluded convincingly that the alcohols are also primary products of the process and that thay are not precursors of the olefins. The authors indicated though that there might be a common precursor for linear α-olefins and linear alcohols.

9.2.2 Secondary Reactions of the α-Olefins

The availability of very efficient gas chromatographic methods (capillary columns, radio gas chromatography) has permitted a detailed determination of the many isomers contained in the synthesis products at higher conversions and hence provided an insight into the course of the secondary reactions. Pichler and Krüger (1973) reported a complete analysis of the C_7 hydrocarbon fractions produced in Sasol's fixed-bed and fluid-bed processes (see Table 9.3). The products are predominantly linear, especially with the fixed-bed catalyst. The non-linear part is > 90% monomethyl-, 6−7% dimethyl-, and 2−3% ethyl branched. The distribution of alkanes and alkenes, as well as the presence of inner double bonds, indicate that the main secondary reaction of the primary α-olefins is hydrogenation, with some double bond migration also present.

Another important contribution towards the mechanism of the secondary reactions of α-olefins was made by Schulz et al. (1970): ^{14}C-tagged ethylene, propylene, 1-butene, or 1-hexadecene were added to the synthesis gas, and the fate of these α-olefins was monitored by radio gas chromatography. The normal pressure cobalt catalyst system, as well as the medium pressure iron catalyst system were investigated. With cobalt catalysts, over 90% of the added olefin reacts. With iron systems the reaction is somewhat slower. Most of the olefin is converted to alkane by hydrogenation.

The fraction of tagged olefin transformed into alkane is comparable with the fraction of the corresponding FT product with the same carbon number found as paraffin. This is considered as an additional proof for the FT synthesis to result primarily in α-olefins which are then hydrogenated in a secondary reaction.

Table 9.3. Composition of the C_7 hydrocarbon fractions of Sasol's fixed-bed and fluid-bed processes (weight %). According to Pichler and Krüger, 1973.

Compound	Fixed-bed	Fluid-bed
n-Heptane	46.3	13.3
2-Methylhexane / 3-Methylhexene-1	2.1	6.7
3-Methylhexane / 5-Methylhexene-1	2.6	7.7
2,3-Dimethylpentane	0.3	0.3
3-Ethylpentane	0.2	0.7
n-Heptene-1	28.3	36.9
cis-n-Heptene-2	6.7	5.4
trans-n-Heptene-2	8.7	6.6
cis-n-Heptene-3	0.4	0.8
trans-n-Heptene-3	0.8	0.7
2-Methylhexene-1	0.9	3.8
4-Methylhexene-1	0.8	5.0
3-Methyl-cis-hexene-2 / 2-Methylhexene-2	0.1	0.4
3-Methyl-trans-hexene-2	–	0.4
4-Methyl-cis-hexene-2 / 4-Methyl-trans-hexene-2	0.2	0.6
5-Methyl-cis-hexene-2	0.1	1.5
5-Methyl-trans-hexene-2	0.2	1.2
2,3-Dimethylpentene-2	0.1	0.4
2,4-Dimethylpentene-1	0.05	0.4
3,4-Dimethylpentene-1	–	0.8
Naphthene	0.4	0.8
Toluene	0.1	3.5
Unknown	0.7	2.1
Sum of products	100.0	100.0
Sum of paraffins	49.2	19.2
Sum of olefins	49.6	72.4
Sum of linear hydrocarbons	91.6	63.7
Sum of branched hydrocarbons	7.6	29.9

But the tagged α-olefins were also incorporated into growing chains. With increasing carbon number of the olefin, its incorporation decreases. Some 30% of the ethylene and propylene, but only 6% of the n-hexadecene, are incorporated into the FT synthesis products. (This is in accord with the relative polymerization activity of these olefins with transition metal catalysts and is a consequence of their ability to coordinate to a metal center.) The incorporation of ethylene does not lead to chain branching, that of propylene gives methyl branches.

Simultaneously, α-olefins were split, with breaking of the double bond. Since this splitting occurs in a hydrogen atmosphere, additional methane is formed by this secondary process. In the cobalt system, and with some $0.3-0.8$ vol% of the tagged olefin in the gas feed, $4-9\%$ of the ^{14}C was found as methane.

Interestingly, also [2-^{14}C]-propene gave a considerable amount of tagged methane. Hence, the following decomposition reactions have to be considered:

$$CH_2=CH_2 \longrightarrow 2\,CH_4 \tag{9.2}$$

$$CH_3CH=CH_2 \underset{\searrow\,3\,CH_4}{\overset{\nearrow\,CH_3CH_3+CH_4}{}} \tag{9.3}$$

$$RCH=CH_2 \longrightarrow RCH_3 + CH_4. \tag{9.4}$$

The splitting of the olefins is more important with cobalt catalysts than with the iron systems, and at least part of the high methane content (Fig. 9.1) is probably due to this kind of reactions. The incorporation reaction is also more frequent with the cobalt system; with the iron catalysts hydrogenation predominates.

Neither Co nor Fe produce significant cracking of the paraffins. ^{14}C-tagged butane added to the synthesis gases could be recovered unchanged in over 99% yield. And even an added 2-methylpentadecane, which certainly has a prolonged residence time in the reactor, remained essentially unaltered.

9.2.3 Modified FT Synthesis

The synthesis originally aimed at gasoline and motor oil, but certain modifications of catalyst and/or reaction conditions led to a broad variety of products, the formation of which is closely related to the FT synthesis proper.

Using a *ruthenium catalyst*, and working at high pressure (1000–2000 atm) and temperature (140–200 °C), Pichler and Buffleb (1940), could orient the synthesis to high melting linear paraffins having molecular weights up to 10^6; the "polymethylenes" obtained are essentially identical with Ziegler poly-ethylene. This and other interesting particularities of ruthenium catalysts will be further discussed in Sect. 9.7.

As mentioned above, alcohols and other oxygenated compounds are always found among the FT products in more or less quantity, and sometimes unwanted. From the old "Synthol" work of Fischer and his coworkers it was known that high pressure favors the formation of oxygenated compounds. More recently, interest appears to have shifted back towards the original "Synthol" process. High pressure processes have been reported to give good yields of aliphatic alcohols in the C_1-C_{18} range, important raw materials for detergents, plasticizers, etc. (Schlesinger et al., 1954; DBP 1961). In Soviet patents (1973), *iron catalysts* have been used at temperatures in the range 160–220 °C, pressures of 50–300 atm, and $H_2/CO = 5$ to 20 to produce liquid, oxygen containing compounds in the range C_1-C_{20}, mostly alcohols. Kagan et al. (1966) claimed that at high space velocity selectivities up to 86%

(expressed as percentage of alcohols in total liquid product boiling below 160 °C) can be obtained.

Another modification involves the addition of *ammonia* to the synthesis gas. Ruhrchemie (DBP 1949) reported the preparation of primary amines, in the presence of 0.5−5% of NH_3 in the gas feed. Iron catalysts were used under conditions which, in the absence of ammonia, would result in hydrocarbons. W. R. Grace and Co., more recently, also claimed Fe as the main component of the catalyst for a process of preparing linear aliphatic, primary amines from mixtures of CO, H_2 and NH_3 (USP, 1973). In this patent it was noted that an increasing amount of ammonia reduces the average molecular weight of the product amine, and that no other nitrogen containing compounds are formed in the process. Evidently, the ammonia interferes with the FT chain growth reaction, terminating the molecules.

An extreme case of molecular weight reduction by NH_3 is given by the formation of acetonitrile from CO, H_2 and NH_3 (Henrici-Olivé and Olivé, 1979). Larger amounts of ammonia were added to the synthesis gas; a *molybdenum oxide/SiO₂ catalyst* was used in a flow reactor operated at 400−500 °C and normal pressure. About 15% of the CO was transformed to CH_3CN. The major byproducts were CO_2 and HCN; only small amounts of CH_4 and propionitrile were found. *Iron* and *tungsten* catalysts also gave CH_3CN, although with decreasing activity in this order. In the absence of ammonia the major product was CH_4 (15−20% with the molybdenum catalyst under comparable conditions), with only 1−2% of C_2 and traces of higher hydrocarbons.

Other catalytic reactions of CO + H_2 closely related to the FT synthesis have been treated in the preceding Chapters: the formation of methane (actually the oldest known hydrocarbon synthesis, found by Sabatier and Senderens, 1902) in Chap. 7, and the technically important production of methanol in Chap. 8.

9.3 Distribution of Molecular Weights

9.3.1 The Schulz-Flory Distribution Function

G. V. Schulz (1935, 1936) has derived an equation for the distribution of molecular weights of polymers obtained by a free radical polymerization process, i.e. through a one-by-one addition of monomer to a growing chain. This distribution function is generally applicable if there is a constant probability of chain growth, α, and $\alpha < 1$ (i.e. some reaction delimiting the chain growth is present).

These conditions can safely be assumed to be given during the FT synthesis. The reaction can be operated over long periods of time at a constant rate. While this constant rate indicates a constant number of catalytic sites, the fact that a large number of molecules is formed per active site suggests that a chain transfer reaction takes place, in the course of which a product molecule leaves the active site, and a new chain is started at the same center. Since α-olefins, as well as alcohols, form the primary products of the FT synthesis, we have to consider more than one particular chain transfer reaction. The probability of chain growth is then defined as

$$\alpha = \frac{r_p}{r_p + \Sigma\, r_{tr}} \tag{9.5}$$

where r_p, r_{tr} are the rates of chain propagation and chain transfer, respectively; α is assumed to be independent of the chain length.

The statistic derivation of the distribution function according to G. V. Schulz is lucid and straightforward. The probability for a chain growth step is α. The probability for the growth step to take place P times without interruption is

$$p_P = \alpha_1 \times \alpha_2 \times \alpha_3 \times \alpha_4 \times \ldots \alpha_P = \alpha^P. \tag{9.6}$$

The number of molecules of degree of polymerization P, n_P, is proportional to the probability of their formation

$$n_P = \text{const} \times \alpha^P. \tag{9.7}$$

The mass (weight) fraction m_P is proportional to n_P, as well as to the molecular weight of the molecules under consideration, $M_P = M_M \times P$ (M_M = molecular weight of the monomer); hence

$$m_P = A \times P \times \alpha^P \tag{9.8}$$

where A contains the constant M_M. The mass fraction is defined in such a way that the sum of all m_P is unity. Moreover, the mass fraction is considered to be a continuous function of P. (This last statement is perfectly admissible for a large average molecular weight; its consequences for small average molecular weight will be discussed below.) It follows then:

$$\int_0^\infty m_P \, dP = A \int_0^\infty P\, \alpha^P \, dP = 1$$

and

$$A = 1/\int_0^\infty P\,\alpha^P\,dP.\qquad(9.9)$$

Solving the integral (taking into account that $\alpha < 1$, $\alpha^\infty = 0$), and combining Eqs. (9.8) and (9.9), leads to the mass distribution function

$$m_P = (\ln^2 \alpha)\,P\,\alpha^P.\qquad(9.10)$$

For practical uses it is conveniently written in the logarithmic form:

$$\log \frac{m_P}{P} = \log(\ln^2 \alpha) + (\log \alpha)\,P.\qquad(9.11)$$

If a molecular weight distribution follows this law, a plot of $\log(m_P/P)$ versus P should result in a straight line. Logarithmic representations of experimental data are often regarded as somewhat less accurate, smoothing any errors and deviations. But this particular one has an internal control, since α is contained in the slope ($\log \alpha$), as well as in the intercept of the straight line with the ordinate [$\log(\ln^2 \alpha)$], and the values of α from both sources have to agree. Figure 9.2 shows a series of theoretical straight lines, in the range of α and P interesting for our present purpose. It should be noted that the anatomy of Eq. (9.11) is such that the slope will result in the correct value of α (and hence of the average molecular weight, vide infra), even if only a few single points (mass fractions m_P) are known with sufficient accuracy. The value derived from the intercept, however, will be in error, if the experimental distribution is incomplete.

According to G. V. Schulz, the average degree of polymerization (number average, P_n) is related to the mass distribution function as follows:

$$P_n = 1/\int_0^\infty (m_P/P)\,dP.\qquad(9.12)$$

With the application of Eq. (9.10) this leads to:

$$P_n = 1/\ln^2 \alpha \int_0^\infty \alpha^P\,dP = -\,1/\ln \alpha.\qquad(9.13)$$

P. J. Flory (1936) published the theoretical distribution function for a different type of macromolecule formation, the polycondensation of bifunc-

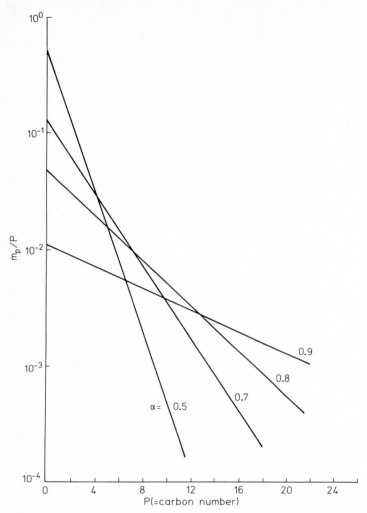

Fig. 9.2. Theoretical Schulz-Flory plots (according to eq. 9.11), for α-values and P range relevant for the FT synthesis

Table 9.4. Comparison of the molecular weight distribution $m_P = f(\alpha)\, P\, \alpha^P$ and of the average degree of polymerization P_n, as given by Schulz (1935, 1936) and Flory (1936).

$\alpha\,(\alpha')$	Schulz $f(\alpha) = \ln^2 \alpha$	Flory $f(\alpha') = (1 - \alpha')^2/\alpha'$	Schulz $P_n = -\,1/\ln \alpha$	Flory $P_n = 1/1 - \alpha'$
0.99	1.01×10^{-4}	1.01×10^{-4}	99.4	100
0.9	0.0111	0.0111	9.5	10
0.8	0.0498	0.0500	4.5	5
0.7	0.1272	0.1286	2.8	3.3
0.5	0.4805	0.500	1.4	2
0.3	1.449	1.633	0.8	1.4

tional monomers, which takes place via the gradual growing together of x-mers (dimers, trimers, etc.):

x-mer + y-mer → (x + y)-mer

with x, y ≥ 1. Flory stated that, although based on an entirely different set of conditions, his equation is essentially equivalent to Eq. (9.10) derived by G. V. Schulz. Following Flory's argumentation, but using, for the purpose of easy comparison, similar symbols as in the derivation of Eq. (9.10), α' is defined as the fraction of functional groups that have reacted at a given time:

$$\alpha' = (N_0 - N_t)/N_0$$

(N_0, N_t = number of molecules present at the beginning and at the time t, respectively). This defines α' also as the probability that a condensation reaction has taken place at a given end group.

For the degree of polymerization P to be realized, the condensation reaction must have taken place (P − 1) times. The probability that no condensation has taken place at both ends is $(1 - \alpha')^2$. Hence the probability of existence of each particular configuration is $\alpha'^{P-1} (1 - \alpha')^2$. The probability that any of the P configurations exists is $P \alpha'^{P-1} (1 - \alpha')^2$, and this is equal to the "weight fraction distribution":

$$m_P = \frac{(1 - \alpha')^2}{\alpha'} P \alpha'^P. \qquad (9.14)$$

In this form, $(1 - \alpha')^2/\alpha'$ corresponds to $\ln^2 \alpha$ in Eq. (9.10). For α (or α') → 1, Eqs. (9.10) and (9.14) are essentially equivalent; for smaller values of α (or α') there is a slight discrepancy, see Table 9.4.

According to Flory's derivations, the average degree of polymerization (number average) is given by

$$P_n = \frac{1}{1 - \alpha'}. \qquad (9.15)$$

For α (or α') → 1, Eqs. (9.13) and (9.15) are, again, essentially equivalent. For smaller α (or α'), P_n according to Eq. (9.15) is generally somewhat higher; calculated values for P_n are included in Table 9.4. In the lower range of α, Eq. (9.15) may be more appropriate than Eq. (9.13), since the method of integration over the range from 0 to ∞, though an excellent approximation for large degrees of polymerization, is less appropriate when applied to relatively small molecules, where the step-by-step growth mechanism has a greater bearing.

Because of the close overall resemblance of the two mass distribution functions, it has become customary to call them, in either form, "Schulz-Flory distribution function".

Some 15 years later, Friedel and Anderson (1950), based on earlier work of Herington (1946), developed an equation for the products of the Fischer-Tropsch synthesis. Since again the same statistics are involved (as long as branching can be neglected), the resulting equation is equivalent to the former two (although apparently the authors were not aware of the work of Schulz and of Flory).

For the sake of completeness we want to mention briefly what kind of a molecular weight distribution is to be expected if $\alpha = 1$, i.e. if there is neither chain termination nor transfer. (Under certain conditions this type of distribution appears to be approximated in Fischer-Tropsch synthesis, see Sect. 9.7.2.) For the ideal case that all molecules have started the growth at the same time, Flory (1940) has shown that the molecular weight distribution is given by the Poisson function

$$m_P = \frac{e^{-v}\, v^{(P-1)}\, P}{(P-1)!\,(v+1)} \qquad\qquad (9.16)$$

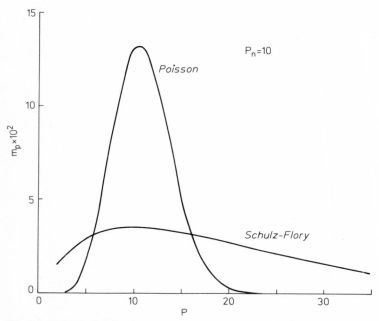

Fig. 9.3. Poisson distribution, according to eq. (9.16), and Schulz-Flory distribution, according to eq. (9.10), for the same average degree of polymerization, $P_n = 10$

where v is the average number of growth steps per molecule which is related to the average degree of polymerization P_n by

$$v = P_n - 1. \tag{9.17}$$

For comparison, the theoretical Schulz-Flory distribution and Poisson distribution, corresponding to the same average degree of polymerization, $P_n = 10$, are given in Fig. 9.3 (Eq. (9.10) with $\alpha = 0.905$ and Eq. (9.16) with $v = 9$, respectively). Evidently, the Poisson distribution is considerably more narrow.

9.3.2 Experimental Molecular Weight Distributions

As mentioned in Sect. 9.1 the products of the FT synthesis vary considerably with catalyst, reaction conditions, (pressure, temperature), and process design. Figs. 9.4–9.7 show that, despite such variations, the Schulz-Flory molecular weight distribution function generally holds.

Figures 9.4 and 9.5 refer to Fischer-Tropsch syntheses oriented towards the production of hydrocarbons. Since the primary α-olefins are partly transformed into inner olefins (isomerization) and paraffins (hydrogenation), the weight fraction m_P for each degree of polymerization is the sum of olefins and paraffins of the same number of carbons. Fig. 9.4 is a plot of the data of

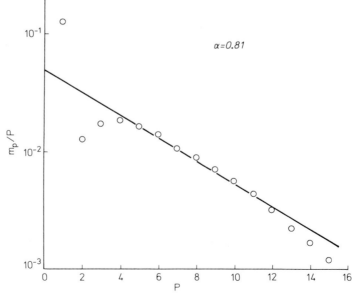

Fig. 9.4. Hydrocarbon data of Fig. 9.1 a represented according to eq. (9.11)

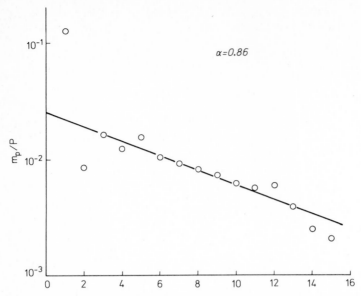

Fig. 9.5. Molecular weight distribution of hydrocarbons, represented according to eq. (9.11). Cobalt catalyst (Data reported by Storch et al., 1951, Chap. 6)

Fig. 9.1 according to Eq. (9.11). In the range C_4-C_{12}, the experimental points lie on a straight line. C_1 is too high and C_2 and C_3 are too low, as are the values beyond C_{13}. The value of α is 0.81 from the slope and 0.80 from the intercept.

Figure 9.5 shows the same plot for data reported by Storch et al. (1951). The range C_3-C_{13} gives a reasonable straight line. Again here the value of C_1 is too high and the values of C_2 and beyond C_{13} are too low. From the slope a value of $\alpha = 0.86$ is obtained; from the intercept $\alpha = 0.86$.

The reasons for the deviations in the range C_1-C_3 have already been mentioned in Sect. 9.2.2: the α-olefins with the greatest ability to be coordinatively bonded to a transition metal (ethylene > propylene) are partly lost by decomposition to methane and, to a lesser extent, by incorporation into growing chains. A certain amount of methane may also proceed from the hydrogenation of carbide, in particular in the cobalt case (cf. Chap. 7). At the upper end of the distribution, the hydrocarbons tend to have longer residence times on the catalyst due to their lower mobility. Presumably, they are partly polymerized, partly cracked, or even carbonized. An additional uncertainty in this range may be caused by the fact that the high boiling fractions give very broad peaks on gas chromatography; their values tend to be underestimated.

Data from the Sasol fluid-bed process, as reported by Pichler and Krüger (1973) are represented according to Eq. (9.11) in Fig. 9.6. Only the weight fractions of the hydrocarbons C_1-C_4 are given explicitly in the reference ($C_1 = 0.131$; $C_2 = 0.101$; $C_3 = 0.162$; $C_4 = 0.132$); the hydrocarbons $\geq C_5$ are grouped together (39 weight %). The accuracy of the weight fractions of the lower hydrocarbons is reflected in the fact that α from the slope as well as from the intercept is 0.64. As mentioned in Sect. 9.1, Sasol's fluid-bed process with iron catalysts leads to a relatively low molecular weight product. From the α value, an average degree of polymerization of $P_n \simeq 2.6$ can be estimated.

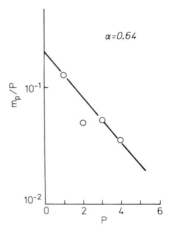

Fig. 9.6. Data from the Sasol fluid-bed process, represented according to eq. (9.11)

Yang et al. (1979) have reported a FT system leading to an even lower average degree of polymerization. The catalyst is coprecipitated $Co/Cu/Al_2O_3$; H_2 and CO are applied at medium pressure, at a ratio ranging from 3:1 to 1:1; the temperature is 225−275 °C. Figure 9.7 shows data for one particular set of conditions (not identified in the reference). The value of α is 0.55 from the slope and 0.52 from the intercept, resulting in an estimated average degree of polymerization of $P_n \simeq 2.2$.

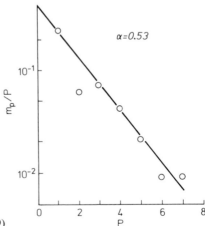

Fig. 9.7. Low molecular weight hydrocarbons, as obtained on a $Co/Cu/Al_2O_3$ catalyst, represented according to eq. (9.11) (according to Yang et al., 1979)

As mentioned in Sect. 9.2.3, the Fischer-Tropsch synthesis can be oriented to give predominantly alcohols, by proper selection of catalyst and reaction conditions. Using data of Kagan et al. (1966), we applied Eq. (9.11) also to alcohols. Figure 9.8 shows two sets of data, obtained with different catalysts. For the Fe_3O_4 catalyst modified with Al_2O_3 and K_2O (Fig. 9.8 a) an average value of $\alpha = 0.37$ has been calculated from the slope (0.38) and from the intercept (0.36). Data of the pure Fe_3O_4 system (Fig. 9.8 b) give the same value $\alpha = 0.54$ from slope and intercept.

From a mechanistic point of view it is of particular interest that amines produced by a modified FT process (cf. Sect. 9.2.3) present the same type of molecular weight distribution. This is shown in Fig. 9.9. In the range C_4-C_{15} the points fit the expected straight line reasonably well. The values of α obtained from the slope ($\alpha = 0.68$) and from the intercept ($\alpha = 0.70$) are in satisfactory agreement; the estimated average degree of polymerization is $P_n \simeq 3.2$.

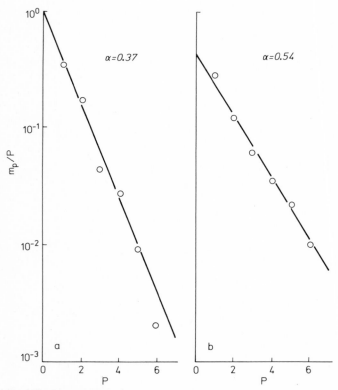

Fig. 9.8 a, b. Molecular weight distribution of alcohols, represented according to eq. (9.11); a) $Fe_3O_4 + Al_2O_3 + K_2O$; b) Fe_3O_4. Data reported by (Kagan et al., 1966)

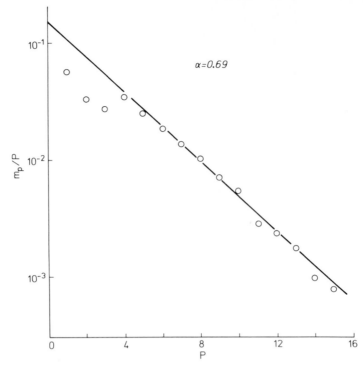

Fig. 9.9. Modified FT process for the production of amines; molecular weight distribution of amines according to eq. (9.11) (Henrici-Olivé and Olivé, 1978)

9.4 Kinetics and Thermodynamics of the FT Reaction

Most kinetic studies published over the years are based on actual process conditions. As Dry et al. (1972) pointed out, this makes interpretations difficult, mainly due to continuous changes of partial gas pressures along the catalyst bed. Using a specially constructed small differential reactor, Dry et al. determined the basic rate law for the FT reaction on an iron catalyst at 225–265 °C. Maintaining the CO partial pressure constant, the H_2 partial pressure was varied, correcting at the same time the space velocity for constant contact time with the catalyst (i.e. working at a constant linear velocity). A clear first order dependence of the overall rate on the H_2 partial pressure was found. The corresponding measurements at constant H_2 and varying CO pressure resulted in zero order dependence of the rate on the latter. Hence

$$\text{rate} = K\, p_{H_2} \tag{9.18}$$

where p_{H_2} is the partial pressure of hydrogen, and K is a constant. This simple rate law has been found valid in the pressure range from 10 to 20 atm, and with a synthesis gas ratio (H_2/CO) varying from 1 to 7. From the temperature dependence of the reaction rate at a given set of conditions an overall activation energy of $\simeq 70$ kJ/mol was determined.

Under actual process conditions the macrokinetics are more complicated. A large body of kinetic data has been collected by Storch et al. (1951). Rate equations featuring a reciprocal proportionality with p_{CO} have been reported by Brötz and Rottig (1952). At temperatures above 300 °C, the water gas shift reaction becomes important; water vapor has been found to depress the rate (Dry et al., 1972). Overall activation energies in the range 84−113 kJ/mol have been reported by Pichler and Krüger (1973) for cobalt catalysts.

There is a general consensus, that the reaction rate is amazingly low, as compared with that of other reactions in the field of heterogeneous catalysis (e.g. Pichler and Krüger, 1973; Vannice, 1975; Dautzenberg et al., 1977). Dautzenberg et al. (1977) have undertaken the task to find out whether the low rate is due to the fact that only very few exposed surface metal atoms are active, or the active sites themselves have a very low intrinsic activity. The authors used a pulse technique, in the course of which a catalyst (ruthenium) was repeatedly exposed to a CO/H_2 mixture at 210 °C and 10 atm during a variable time τ. In between exposures, the system was quenched by flushing with hydrogen and heated to 350 °C for forced chain termination and product release. It was found that the average chain length, as well as the product molecular weight distribution, depend sensitively on the length of τ. For instance for $\tau = 8$ min, the C_{12}/C_6 molar ratio was 0.12, whereas under stationary conditions the same system would have resulted in a value of 0.74 for this ratio. On increasing the pulse time, the production of long chain hydrocarbons was enhanced, and the relative distribution of the lighter hydrocarbons approached that obtained under steady state conditions. Moreover, the fact that the lighter hydrocarbons approach their steady state situation at τ values where the long chain hydrocarbons are still in considerable deficiency indicates that the first growth step is not essentially slower than the following ones. This work does not answer the question which partial step in the course of the complex transformation from CO to a $-CH_2-$ group is actually rate determining. However, the experimental rate law (Eq. 9.18) indicates that the rate determining step involves hydrogen. Based on an elaborate kinetic model, the authors were able to estimate a formal rate constant of actual chain growth, $k_{propagation} \simeq 1.5 \times 10^{-2}$ sec^{-1}, corresponding to a growth rate of ca. one $-CH_2-$ group per minute per growing chain. This work appears to relegate the cause of the low overall rates to the slow chain growth rather than to an extremely small number of active sites, at least for the case of ruthenium catalysts under the prevailing conditions. On the other hand Biloen et al. (1983), using a similar transient method, estimated the fraction of surface metal atoms covered with active intermediates for unsupported Co, Ni/SiO$_2$ and Ru/Al$_2$O$_3$ catalysts, and came up with values varying from $\simeq 10\%$ for cobalt to 10−15% for Ni and Ru.

Most workers appear to be aware by now that the number of active sites cannot be equalized to the number of surface metal atoms, as determined by CO, H_2 or O_2 adsorption. The reasons are manifold. Yates et al. (1979) have shown that more than one CO can be coordinated to a surface metal atom (cf. Table 3.6); the same may be true for hydride ligands. Hydrogen spillover (Sect. 2.5) as well as SMSI effects (Sect. 6.2.2) have proven that hydrogen migrates into the support. Moreover, there is increasing evidence that only metal atoms at especially exposed sites (steps, kinks), where the availability of coordination sites is relatively high, are catalytically active (see the discussion of the SMSI effect in Sect. 6.2.2, as well as Fig. 7.3 and the concomitant text). The uncertainty of the number of active sites makes it necessary to apply indirect methods (as done by Dautzenberg et al.) if absolute values for the rate constants are to be determined.

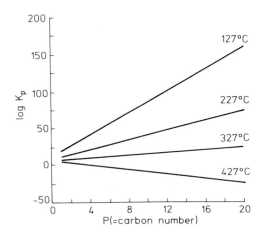

Fig. 9.10. Equilibrium constant (K_p) as a function of the chain length of the product molecules, and temperature (according to Pichler and Krüger, 1973)

The synthesis of hydrocarbons from CO and H_2 is a strongly exothermic reaction, evolving some 146−176 kJ per mole of CO converted under the usual reaction conditions (Storch et al., 1951; Pichler and Krüger, 1973). Since the product distribution depends sensitively on the reaction temperature, heat removal is a very important factor in process design.

The equilibrium constants for the formation of hydrocarbons of varying chain length depend not only on the temperature, but also on this length. Figure 9.10 shows that the yield of higher hydrocarbons can be expected to decrease with increasing temperature. Below 400 °C, the formation of the higher hydrocarbons is favored as compared to the lower ones, since the equilibrium constant increases with the carbon number. However, a graph such as that in Fig. 9.10 can only indicate general trends. The actual FT synthesis is governed by a large number of parallel primary reactions (e.g. formation of hydrocarbons and alcohols), secondary reactions (e.g. hydrogenation and isomerization), and side reactions (e.g. water gas shift equilibrium, disproportionation of CO to C and CO_2), resulting in a complex system of simultaneous equilibria which resists exact evaluation (Anderson and Emmett, 1956).

9.5 Reaction Mechanism

9.5.1 Suggested Mechanisms

The mechanism of the Fischer-Tropsch synthesis has attracted much interest in the past decade. Some authors favor the original proposal made by Fischer and Tropsch (1926), according to which the synthesis proceeds via the formation of carbides. They assumed that the finely divided, carbon-rich carbides are decomposed by hydrogen with regeneration of the catalyst metal, and formation of the hydrocarbons. Evidently, this would imply the formation of CH_2 entities arranged in a row on the surface of the catalyst, unless one assumes that the CH_2, or in general CH_x, fragments move more or less freely along the catalyst surface. (For a review see Muetterties and Stein, 1979.) The essential features of this scheme are given in Eq. (9.19).

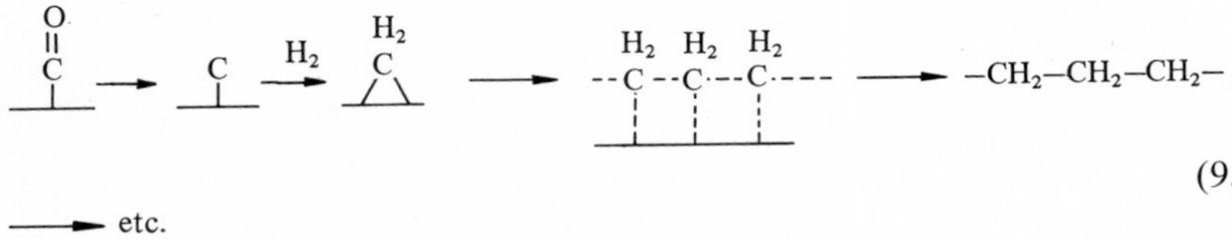

$$\tag{9.19}$$

As early as 1958 it was suggested that the chain growth in the FT synthesis might involve the insertion of CO in metal-carbon bonds (Wender et al., 1958). However, this idea was not further pursued, until Pichler and Schulz (1970), as well as Henrici-Olivé and Olivé (1976a) proposed detailed reaction mechanisms involving CO insertion into metal−H bonds (initiation) and metal-

Storch et al. (1951) suggested the formation of hydroxymethylene groups, from carbon monoxide chemisorbed at the metal surface, and hydrogen chemisorbed in atomic form; C−C bonds are then to be established through a condensation reaction between the hydroxymethylene groups, with loss of H_2O. Here, again, a perfect alignment of the hydroxymethylene groups in a row is implicit, unless one assumes a free movement of carbenes on the surface. The basics of this mechanism are summarized in Eq. (9.20).

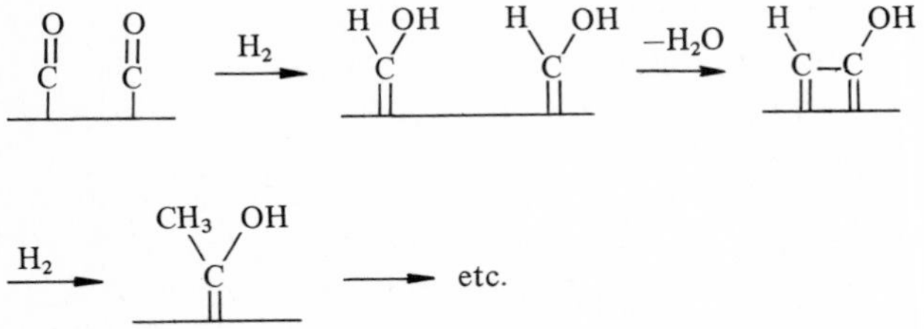

$$\tag{9.20}$$

alkyl bonds (chain growth), analogous to the well known CO insertion into such bonds in homogeneous catalysis, and followed by reduction of the acyl group. The two mechanisms diverge in the subsequent steps, the second being more tightly related to known patterns in coordination and organometallic chemistry. This will be discussed in detail in Sect. 9.5.3. But, without going into details at the moment, the general growth pattern can be summarized as shown in Eq. (9.21):

$$
\begin{array}{ccc}
& \overset{\displaystyle R}{\overset{|}{C}O} & \overset{\displaystyle R}{\overset{|}{C}H_2} \\
\overset{\displaystyle R}{\overset{|}{M(CO)_n}} \to & \overset{|}{M(CO)_{n-1}} & \xrightarrow[+\,H_2/CO]{-\,H_2O} M(CO)_n + H_2O
\end{array} \tag{9.21}
$$

with R = H or alkyl group. The important difference with regard to Eqs. (9.19) and (9.20) resides in the fact that the entire chain growth takes place at one and the same metal center, at the surface of a catalyst. (Pichler's scheme requires a second neigboring metal center to take care of the transformation of the oxygen to water, vide infra.)

9.5.2 The Carbide Theory in the Light of Homogeneous Coordination Chemistry

Ever since the *carbide theory* for the chain growth in the FT synthesis (see Eq. 9.19) appeared, there has been controversy about its validity. After years of arguments and counterarguments, the problem appeared temporarily settled by Kummer et al. (1948). Using ^{14}CO or ^{12}CO for the formation of carbide on the surface of iron and cobalt catalysts, and running the synthesis with $^{12}CO + H_2$ or $^{14}CO + H_2$, respectively, these authors studied the distribution of radioactivity in the initially formed hydrocarbons. They concluded that the carbide route could not account for more than 10% of the product.

After this work, the carbide theory was more or less shelved until the seventies when, motivated by the oil embargo, work on the Fischer-Tropsch synthesis was actively resumed. At that time, the concepts of coordination chemistry had made their way into the interpretation of catalysis, in particular in homogeneous phase, and reaction paths more related to those established for homogeneous catalysis were suggested for the FT (see Eq. 9.21). Moreover, Takeuchi and Katzer (1981) have provided convincing evidence that CO is nondissociatively hydrogenated to methanol (see Sect. 8.1). It should be recalled that olefins and alcohols are the primary products of the FT synthesis (cf. Sect. 9.2.1) and that it has been suggested and supported that they have a common precursor (Henrici-Olivé and Olivé, 1976a). Hence, the findings of Takeuchi and Katzer appear to corroborate the CO insertion mechanism.

Other authors, however, stuck to the carbide theory of chain growth, e.g. Biloen et al., 1979, 1983; Nijs and Jacobs, 1980; Falbe and Frohning, 1982; Kellner and Bell, 1981; Brady and Pettit, 1980, 1981. Also Rofer-de Poorter (1981) appears to favor the "condensation of hydrocarbon intermediates as a carbon-carbon bond forming step", but includes also CO insertion in a compilation of all relevant reactions for which experimental evidence has been reported in the literature. In the following we shall discuss those data which are most frequently cited as evidence for the carbide mechanism.

Brady and Pettit (1980, 1981) studied the behavior of methylene fragments, as produced by the thermal decomposition of diazomethane, on various transition metals. The authors came to the conclusion that their observations are in agreement with the mechanism depicted by Eq. (9.19). Their main findings are a) in the absence of hydrogen and CO, only ethylene is formed; b) in the presence of H_2, hydrocarbons up to C_{18} featuring a Schulz-Flory molecular weight distribution are observed; c) in the presence of H_2/CO, the same type of distribution, but with an average degree of polymerization higher than that for H_2/CO alone, is observed. However, the findings of Brady and Pettit do not contradict the assumption that the FT synthesis takes place by a CO insertion mechanism (Eq. 9.21), as discussed recently by the present authors (Henrici-Olivé and Olivé, 1982a). The polymerization of CH_2 fragments originating from the decomposition of CH_2N_2 by a soluble metal hydride catalyst is described in the literature (Mazzi et al., 1970). Presumably, the CH_2 fragments are inserted into metal$-$H (initiation) and metal-alkyl (chain growth) bonds. In fact the insertion of CH_2 groups originating from the decomposition of diazomethane into a metal$-$H bond of a model complex has been reported by Puddephatt and coworkers (Azam et al., 1982):

$$\left[\begin{array}{c} P \quad P \\ | \quad | \\ H-Pt-Pt-L \\ | \quad | \\ P \quad P \end{array}\right] [PF_6] \xrightarrow{CH_2N_2} \left[\begin{array}{c} P \quad P \\ | \quad CH_2 \; | \\ Pt \quad Pt \\ CH_3 | \qquad | \; L \\ P \qquad P \end{array}\right] [PF_6] \qquad (9.22)$$

($P-P = Ph_2PCH_2PPh_2$; $L = CO$ or PMe_2Ph). At the same time a μ-methylene bridge is also formed. In this particular case, the methyl and methylene groups cannot couple on pyrolysis because they are trans to each other (only methane is formed). But the insertion of a CH_2 ligand into a metal-alkyl bond in cis position has been observed in another model complex (see Eq. 5.24).

Hence, the data of Brady and Pettit are more consistently interpreted by a step-by-step insertion polymerization of CH_2 units on a metal hydride catalyst which, of course, results in Schulz-Flory distribution. In the absence of H_2, the polymerization of CH_2 fragments can not take place because no metal-hydride catalyst is formed. Actually, Brady and Pettit reported that only ethylene (the product of the uncatalyzed dimerization of CH_2 fragments) is formed in the absence of H_2.

The interpretation of Brady and Pettit's results as a hydride initiated insertion polymerization of CH_2 units is corroborated by the fact that Ni and Pd catalysts form higher hydrocarbons from diazomethane-born units. Under the reaction conditions used (normal pressure, $25-200\,°C$, SiO_2 support) these catalysts would not be Fischer-Tropsch active. Ni would require high pressure to produce anything but methane (68 atm as mentioned by Brady and Pettit themselves); Pd produces small amounts of CH_4 and CH_3OH at 1 atm and $215\,°C$ (Driessen et al., 1983), and selectively methanol at higher pressure (Sect. 8.3). Hence, the higher hydrocarbons appear to have grown by a mechanism different from that of Fischer-Tropsch growth.

The increase of the molecular weight, within the framework of a Schulz-Flory distribution, on introducing CH_2N_2 into a Fischer-Tropsch system, appears then as a consequence of a copolymerization with different "monomers", $(CO + H_2)$ and CH_2N_2 which, however, lead to the same incorporated unit, $-CH_2-$. Adapting Eq. (9.5) to this situation, one obtains:

$$\alpha'' = \frac{\Sigma\, r_p}{\Sigma\, r_p + \Sigma\, r_{tr}} \tag{9.23}$$

where $\Sigma\, r_p$ is the sum of the rates of the two growth steps. If $\alpha'' > \alpha$, a Schulz-Flory distribution with a larger average molecular weight results.

Biloen et al. (1979) also advocate the carbide theory of the FT synthesis. the authors carried out experiments similar to those of Kummer et al., mentioned above (except for working with ^{13}C instead of ^{14}C, and using mass analysis), but they came to the opposite conclusion. They suggested that "in the Fischer-Tropsch synthesis CO dissociates in a fast step to give carbidic intermediates, from which both methane and higher hydrocarbons are produced". Their main evidence is the presence of more than one ^{13}C in the lower hydrocarbons, if ^{13}C was deposited onto the surface, and the synthesis was carried out with $^{12}CO + H_2$ at low conversion. The data obtained with Co and Ru catalysts are reproduced in Table 9.5. The authors investigated also Ni

Table 9.5. Isotopic composition of methane and propane (mol%) produced when $^{12}CO + H_2$ was admitted to ^{13}C deposited from ^{13}CO on metal; according to Biloen et al. (1979).

Hydrocarbon	CH_4		C_3H_8			
^{13}C/molecule	0	1	0	1	2	3
Co catalyst[a]	27	73	62	26	10	2
Ru catalyst[a]	70	30	31	27	32	4

[a] Batch reactor; $T = 170\,°C$; $p = 1.3-1.4$ atm; $H_2/CO = 3$; gas volume $(^{12}CO + H_2)$ such that the number of C atoms in the gas phase is not substantially larger than the number of surface exposed metal atoms; analysis of the product after consumption of $\simeq 20\%$ of the deposited ^{13}C.

catalysts; these data parallel those of Sachtler et al. (1979) mentioned in Sect. 7.2, and will not be discussed here further, since Ni is not a typical Fischer-Tropsch catalyst (giving mainly methane at low pressure).

Several points should be borne in mind while evaluating the data of Table 9.5. In the cobalt case, 27% of CH_4 and 62% of C_3H_8 do not contain any ^{13}C. (C_2's are not mentioned.) Moreover, making use of the value of the Schulz-Flory distribution parameter $\alpha \simeq 0.7$ reported by the authors it can easily be computed that the fraction $C_1 + C_3$ represents only 22% of the total hydrocarbons produced. Nothing is reported concerning the presence and distribution of ^{13}C in the hydrocarbons representing 78% of the product. In the ruthenium case ($\alpha \simeq 0.9$), $C_1 + C_3$ represent only 3.4%, hence 96.6% of the formed products are not accounted for. These data, thus, do not provide a solid evidence for the carbide mechanism to account for FT chain growth. However, the fact remains that a significant amount of the methane and small amounts of lower hydrocarbons do actually contain one, two or even three ^{13}C atoms. Evidently, part of these lower hydrocarbons can be formed in a way different from CO insertion. But, do we really have to consider "CH_x particles" moving freely along the catalyst surface until they meet and form hydrocarbons?

It appears that modern coordination and organometallic chemistry offers more probable reaction paths. The insertion of terminal (methylidene) or bridging (μ-methylene) carbene groups into metal-alkyl bonds, within one and the same complex or cluster, has repeatedly been reported (see Sect. 5.3.6). In view of these interesting reactions, an occasional insertion of a carbene group into a growing FT chain appears quite probable.

A number of other observations may be relevant (Henrici-Olivé and Olivé, 1983, and references therein). The dimerization of carbene groups is known, but whenever it has been observed, it took place within the confines of a binuclear complex or a cluster (presumably with simultaneous bond breaking and bond making). A certain mobility of carbene ligands within the confines of a cluster or a bimetallic intermediate, involving an exchange between terminal and bridging positions has also been reported. Moreover, it should be remembered that the transfer of alkyl ligands from one metal to another is common practice in preparative metallorganic chemistry; here again the pathway is assumed through intermediate binuclear complexes. Such internuclear alkyl group exchange is best documented for the group III metal trialkyl compounds, but there is also increasing evidence that it takes place with transition metal compounds. Finally rapid intranuclear motion of hydride ligands within metal clusters has amply been demonstrated by NMR data; the activation energy for such motion may be as low as $12-20$ kJ/mol (see Sect. 2.5).

In the light of this knowledge from homogeneous coordination and organometallic chemistry, the formation of methane and some lower hydrocarbons from carbidic carbon, as observed by Biloen et al., may be visualized as follows: Carbide-carbon is generally located in the interstices between metal atoms, at equal distance to several metal atoms (see e.g. Fig. 3.8). If an oxidative addition of hydrogen takes place at any of the surrounding metal atoms, carbene ligands can be formed. These carbene ligands may have a certain mobility, the movement taking place by changes from bridging to terminal positions or viceversa. Certainly, the carbene ligand remains at all times coordinatively bonded to at least one metal atom. The driving force for such movement would be, presumably, an energy gain through more favorable ligand environment at the acceptor metal center. The same restricted mobility may also be assumed for the alkyl ligands (growing chains in the Fischer-Tropsch synthesis), albeit only on the surface.

The most probable fate of the carbene ligand is the capture of two more hydrogen atoms to give methane. The small hydrogen, either as molecule or, more probably, dissociated, can be assumed to have the easiest access to carbene ligands, even to those located in the inner part of the catalyst particles. In fact, the major part of the labeled carbide carbon ends up as methane, as shown by Biloen et al. It should also be noted that methane is frequently found in large excess among the reaction products of the FT synthesis, as compared to the otherwise "normal" (Schulz-Flory) distribution of molecular weights (see e.g. Figs. 9.4 and 9.5). Presumably, there is always some methane formation via the carbide-carbene route, parallel to the Fischer-Tropsch chain growth to higher hydrocarbons.

If a carbene ligand and an alkyl group (growing chain) happen to be simultaneously coordinated to one and the same metal center, the first may be inserted into the metal-alkyl bond. For steric reasons, this may be possible only at the catalytically active surface metal centers.

An encounter of two methylene groups (at two neighboring metal centers, which must not necessarily be Fischer-Tropsch active) evidently may lead to ethylene formation. However, ethylene has a large tendency to coordinate to a metal center, instead of escaping into the gas phase; as a consequence it may become incorporated into growing chains, as has been shown by Schulz et al. (1970), with the aid of tagged ethylene. Hence, this is another pathway for carbide carbon to get into the hydrocarbons.

The formation of C_3 hydrocarbons from carbide carbon alone should have a very low probability in a Fischer-Tropsch system. In principle, such C_3 may proceed from the insertion of carbide-based ethylene into a likewise carbide-based metal-methyl or metal-methylidene bond. In the latter case, cyclopropane would probably be formed first, as shown by Casey with model complexes (Casey, 1979), but in the presence of hydrogen it would end up as propane (Merta and Ponec, 1969). However, in the inner parts of a catalyst particle, the steric conditions are most probably not given for such chain prolongation. At the catalyst surface, on the other hand, the reaction of any carbide-based intermediate with a growing chain appears more probable. Actually, only very small amounts of molecules with three carbide-based carbon atoms (and none with more than three) have been claimed by Biloen et al.

Summarizing, the presence of some tagged, carbide-based carbon in the Fischer-Tropsch products cannot be considered as strong evidence for the carbide mechanism of Fischer-Tropsch chain growth.

A few words may be appropriate concerning the mechanism suggested by Storch et al. (Eq. 9.20). Apparently, this mechanism is not considered probable any longer, primarily because no evidence for M=C(R)OH species has been found on surfaces, and because condensation reactions of that type have no precedent in model complexes. (For a discussion see Rofer-de Poorter, 1981.)

Although the carbide route is not considered as a major pathway to higher hydrocarbons, the formation of carbides on FT catalysts, in particular Fe catalysts, is a reality. At least four different iron carbides have been identified, whereby Mössbauer spectroscopy was of great help (Raupp and Delgass, 1979; Niemantsverdriet et al., 1980; Schäfer-Stahl, 1980). However, no clear relationship between the amount of any of the carbides, or the amount of remaining metallic Fe, on one side, and the catalytic activity during the steady state FT synthesis on the other, has been detected. Apparently, the relatively small amount of active sites is independent of this variable. More important may be the availability of coordination sites, i.e. the position of the active sites at especially exposed locations (steps, kinks, growth spirals).

9.5.3 Details and Support for the CO Insertion Mechanism

Any plausible reaction scheme for the FT synthesis has to take into account the experimental facts that α-olefins as well as alcohols are the primary products, thus precluding one as the precursor of the other (Sect. 9.2.1). It has also to consider that the primary products have a Schulz-Flory molecular weight distribution.

The present authors have suggested the reaction mechanism shown in Scheme 9.1 (Henrici-Olivé and Olivé, 1976a). At that time, it was based more on chemical intuition than on experimental proof of every individual step. In formulating the Scheme, we followed a contemporary trend, probably first expressed by Nyholm (1965), of looking at the heterogeneous catalysis with transition metals more from the point of view of individual active centers and of their coordination chemistry, than that of "active surfaces". Therefore, the Scheme was based, as far as possible, on individual steps well established in the homogeneous catalysis with soluble transition metal complexes. Such individual steps are, e.g., coordination of CO and of olefins, oxidative addition and reductive elimination, insertion, as well as β-H transfer (see Chap. 5). In the meantime, a large number of papers by various authors has appeared confirming most of the intermediates assumed in this Scheme, either directly, or by analogous reactions.

The first part of Scheme 9.1 (up to step 8) has been amply discussed in the Chapter on methane (Sect. 7.2, Scheme 7.1). Supporting evidence for this part of the Scheme has been given there, and the reader is referred to that discussion.

With step 9 in Scheme 9.1 the reaction cycle begins again, the difference being that the carbon monoxide is now inserted into a metal-alkyl bond instead of a metal−H bond. The insertion of coordinated CO into a metal-

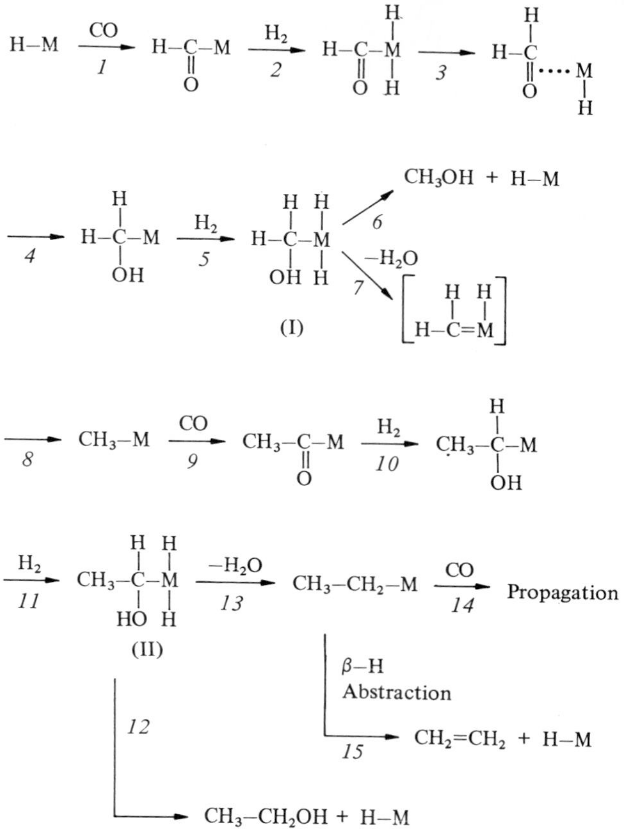

$$H-M \xrightarrow[1]{CO} H-\overset{\displaystyle ||}{\underset{\displaystyle O}{C}}-M \xrightarrow[2]{H_2} H-\overset{\displaystyle H}{\underset{\displaystyle O}{\underset{\displaystyle |}{C}}}-\overset{\displaystyle |}{M} \xrightarrow{3} H-\overset{\displaystyle H}{\underset{\displaystyle O}{C}} \overset{\displaystyle ||}{\cdots} \overset{\displaystyle M}{\underset{\displaystyle H}{|}}$$

$$\xrightarrow{4} H-\overset{\displaystyle H}{\underset{\displaystyle OH}{\underset{\displaystyle |}{C}}}-M \xrightarrow[5]{H_2} H-\overset{\displaystyle H\ H}{\underset{\displaystyle OH\ H}{C}}-M \quad (I)$$

$$\overset{6}{\nearrow} CH_3OH + H-M$$
$$\overset{7}{\searrow}_{-H_2O} \left[H-\overset{\displaystyle H\ H}{C}=M \right]$$

$$\xrightarrow{8} CH_3-M \xrightarrow[9]{CO} CH_3-\overset{\displaystyle ||}{\underset{\displaystyle O}{C}}-M \xrightarrow[10]{H_2} CH_3-\overset{\displaystyle H}{\underset{\displaystyle OH}{C}}-M$$

$$\xrightarrow[11]{H_2} CH_3-\overset{\displaystyle H\ H}{\underset{\displaystyle HO\ H}{C}}-M \xrightarrow[13]{-H_2O} CH_3-CH_2-M \xrightarrow[14]{CO} \text{Propagation}$$
$$(II)$$

$$\beta-H \text{ Abstraction}$$

$$\overset{15}{\longrightarrow} CH_2=CH_2 + H-M$$

$$\overset{12}{\longrightarrow} CH_3-CH_2OH + H-M$$

Scheme 9.1

alkyl bond present at the same transition metal center is one of the key reactions in homogeneous catalysis. It is an easy reaction, and it is theoretically well understood (see Sect. 5.3.1 and 5.3.2). The next step (step 10) actually summarizes the results of three steps, corresponding to steps 2–4, with the difference that the carbon chain is now increased by one unit. Oxidative addition of H_2 (step 11) leads to the intermediate (II) which can undergo two alternative reactions, giving either the alcohol by reductive elimination (step 12), or the alkyl group by H_2O elimination (step 13, which summarizes the two steps corresponding to steps 7 and 8). The alkyl-metal compound can either add CO and thus contribute to the chain propagation (step 14) or, by β-H transfer, give an α-olefin (ethylene at this stage) and a metal hydride which continues the kinetic chain (step 15).

The only new feature in Scheme 9.1 as compared with Scheme 7.1 is the β-H abstraction (step 15), which can take place as soon as a β-H becomes available, and which leads to the primary products of the FT synthesis, the α-olefins. This reaction is also one of the key reactions in homogeneous catalysis and has been discussed in Sect. 5.4.1. For instance in the dimerization, oligomerization and polymerization of ethylene, β-H abstraction is the molecular weight determining step:

$$RCH_2CH_2-M \rightarrow RCH=CH_2 + M-H. \tag{9.24}$$

From a kinetic point of view it is a chain transfer (not termination) reaction, since the resulting metal hydride is a chain carrier (i.e. it initiates a new chain), and as such it is suggested in Schemes 7.1 and 9.1. The analogy goes further than that. It has been shown for the ethylene polymerization that electron acceptor ligands on the metal favor the β-H abstraction, resulting in a smaller average molecular weight of the products (Henrici-Olivé and Olivé, 1974). On the other hand it is known since the old days of the FT synthesis, that decreasing the ratio metal/kieselguhr in the catalyst leads to a decrease of the average molecular weight of the products (Ullmann, 1957). The carrier kieselguhr may be considered as a highly acidic (electron acceptor) "ligand" of the surface metal atoms.

Scheme 9.1 explains, at least qualitatively, most of the experimental findings without violating known principles concerning the reaction patterns at transition metal centers. It should be noted that at no stage does the Scheme assume atomic metal. Although the catalysts are reduced to the metallic state in the "activating" treatment generally applied to them before use ($\geq 500\,°C$, H_2 and/or CO stream), the active sites are certainly not zerovalent metal atoms under reaction conditions. As soon as hydrogen is oxidatively added to a metal center (chemisorbed in the language of surface scientists), the oxidation number increases by either two or one, depending on the addition mode of the hydrogen molecule, either to one metal center (cf. Eq. 5.1), or to two metal centers (cf. Eqs. 2.2, 5.3). In Scheme 9.1 we start with a metal$-$H bond, having omitted for clarity all other ligands which may be H, CO and/or other metal atoms. From there on the Scheme assumes a continuous alternation of oxidative additions and reductive eliminations, in the same way as it is observed in many well understood homogeneous reactions catalyzed by transition metal complexes (see e.g. Henrici-Olivé and Olivé, 1977).

The reaction scheme suggested by Pichler and Schulz (1970) is in agreement with Scheme 9.1 in the most important feature, i.e. the chain growth proceeding by successive CO insertions at one and the same metal center. It diverges in several of the other steps. Thus, it assumes that hydrogen is activated (chemisorbed) at a metal center different from the one bearing the formyl (acyl) group. Pichler's Scheme also assumes that the intermediate formaldehyde (and higher aldehydes) are μ-bonded to two neighboring metal centers:

$$
\begin{array}{ccc}
 & H & H \\
 & | & | \\
R-C-M+H-M' \rightarrow R-C-O \xrightarrow[-H_2O]{+H_2} R-C=M \\
\| & \quad | \quad | & \\
O & M \ M' &
\end{array}
\tag{9.25}
$$

(R = H or alkyl). In model complexes this type of bonding has been observed thus far for a zirconium complex only, whereas the formaldehyde complexes of the higher transition metals show π-bonding to only one metal center, as suggested in Scheme 9.1 (cf. Table 4.1). No hydroxymethyl (or methoxy) intermediate is formulated (cf. Scheme 8.1, Eq. 8.3, and the consequences thereof). Moreover the formation of the product α-olefins is not assumed to take place by β-H abstraction and reestablishment of the chain carrier (step 15 in Scheme 9.1), but by a rearrangement of the carbene ligand:

$$\begin{array}{c} H \\ | \\ R-CH_2-C=M \rightarrow R-CH=CH_2 \rightarrow R-CH=CH_2 + M. \\ \vdots \\ M \end{array} \qquad (9.26)$$

Thus some aspects of Pichler's Scheme lack organometallic support.

Any suggested mechanism has to accommodate the fact that linear aliphatic primary amines are formed as the only nitrogenated products, if the FT synthesis is carried out in the presence of ammonia, and otherwise under conditions that would lead to linear hydrocarbons in the absence of NH_3. The amines present a Schulz-Flory molecular weight distribution, and the average chain length decreases with increasing amount of NH_3 in the synthesis gas (cf. Sect. 9.2.3 and Fig. 9.9). If the hydrocarbons were to be built up from "carbidic intermediates CH_x", α, ω-diamines should be expected among the reaction products. The CO insertion mechanism, on the other hand, can explain this selectivity. The Schulz-Flory distribution could, in principle, be expected if the primary products of the FT synthesis, the α-olefins, were to react with ammonia in a subsequent step, either outside the coordination sphere of the catalyst, or after coordination to a metal center. However, such subsequent amination could not explain the dependence of the molecular weight of the amines on the ammonia concentration.

The facts are best interpreted assuming NH_3 acting as a chain transfer agent at some stage of the growth cycle, resulting in the amine and a metal hydride (Henrici-Olivé and Olivé, 1978). In the course of the growth cycle (see Scheme 9.1) there are alkyl, acyl, hydroxyalkyl, and carbene groups ligated to the metal center. The alkyl ligand offers the most straightforward route:

$$R-M + NH_3 \rightarrow R-NH_2 + M-H \qquad (9.27)$$

(M = metal center). Amazingly, no reference to this simple type of reaction has been found thus far in the literature.

The acyl ligand might be visualized to react with ammonia giving an imine ligand which, on subsequent hydrogenation, would result in a primary amine and a metal hydride

$$R-\underset{\underset{O}{\|}}{C}-M + NH_3 \xrightarrow{-H_2O} R-\underset{\underset{NH}{\|}}{C}-M \xrightarrow{H_2} RCH_2NH_2 + M-H. \tag{9.28}$$

Markó and Bakos (1974) have shown that aldehydes and ketones are transformed to amines under similar reaction conditions, although primary and secondary amines are formed in comparable amounts if ammonia is used as amination agent; in the modified Fischer-Tropsch system, on the other hand, only primary amines are formed. Moreover, it is felt that the mechanism depicted in Eq. (9.28) should lead to nitrogen containing byproducts such as amides and nitriles, which are not found experimentally.

The OH groups of the α-hydroxylalkyl ligands could also react with ammonia:

$$R-\underset{\underset{OH}{|}}{CH}-M + NH_3 \xrightarrow{-H_2O} R-\underset{\underset{NH_2}{|}}{CH}-M \xrightarrow{+H_2} RCH_2NH_2 + M-H.$$

The catalyzed production of amines from alcohols and ammonia is well known; it requires, however, temperatures of $300-400\,°C$ (Ullmann, 1957b), whereas the modified Fischer-Tropsch synthesis operates at $200\,°C$.

Finally, one might consider the carbene ligands, formulated as the result of step 7 (and corresponding later steps; see Scheme 9.1), to react with ammonia. Fischer and coworkers (Klabunde and Fischer, 1967; Weiss and Fischer, 1976) have actually shown that ammonia is able to react with transition metal-carbene complexes (in particular complexes of chromium), but the result is an aminocarbene complex which is so stable that the reaction can be used to protect amino groups in amino acids during peptide synthesis. Similar results have been reported for platinum-carbene complexes (Chisholm et al., 1975). This brief discussion indicates the simple, one-step amination of the alkyl-metal bond, Eq. (9.27), as the most probable reaction. Evidently, such a process would be competitive with the CO insertion (growth step):

$$
R-M \underset{NH_3}{\overset{CO}{<}} \quad
\begin{array}{l}
\nearrow R-\underset{\underset{O}{\|}}{C}-M \\[2ex]
\searrow RNH_2 + M-H
\end{array}
$$

which easily explains the molecular weight reduction produced by increasing the amount of NH_3 in the feed gas mixture.

9.5.4 The Chain Initiating Step

Assuming that the insertion of CO in a metal−H bond (step 1 in Scheme 9.1) might be difficult, Barneveld and Ponec (1978) suggested that CH_x particles produced by the decomposition of CO with following partial hydrogenation, may start the chains by inserting CO. This is certainly a possibility that cannot be a priori discarded. (See, however, Sect. 5.3.5 regarding theoretical and experimental support for step 1 in the Scheme 9.1.) In terms of Scheme 9.1, the reaction sequence could be started at step 8 by carbene ligands or at step 9 by methyl ligands, if these have become available in the course of the hydrogenation of carbide carbon. Evidently, a $M=CH_2$ or $M-CH_3$ species has no memory whether it proceeds from carbidic carbon or from a CO molecule inserted into a M−H bond with the subsequent steps suggested in Scheme 9.1.

It was also pointed out by several authors that carbide, carbyne and carbene ligands react with CO, and that this may have relevance to the FT synthesis. It should, however, be taken into account that none of these reactions takes place by migratory insertion. Rather the CO molecule is added to the carbon of the carbide, carbyne or carbene ligand resulting in a ketene ligand, as illustrated by the following examples.

The complex $[Et_4N]_2[Fe_6C(CO)_{16}]$, with the carbidic carbon in the center of the octahedral Fe core, declustered on oxidation to give the peripheral butterfly carbide shown in Fig. 9.11. The carbide added CO resulting in a ketene which, in the presence of methanol, transformed to a μ_4-carbomethoxy-methylidine ligand (Bradley, 1979).

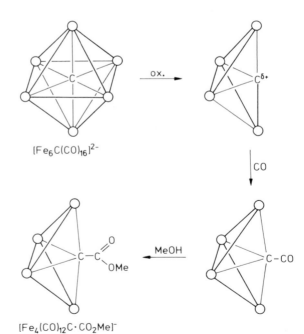

Fig. 9.11. Suggested mechanism for the reaction of $[Et_4N]_2[Fe_6C(CO)_{16}]$ with tropylium bromide (ox) in methanol (Bradley, 1979). Reproduced by permission of the American Chemical Society

$[Fe_6C(CO)_{16}]^{2-}$

$[Fe_4(CO)_{12}C \cdot CO_2Me]^-$

The reaction of model carbyne complexes with CO also led to ketene structures. Kreissel et al. (1978) found the reaction

$$(\eta\text{-}C_5H_5)(CO)[P(CH_3)_3]M \equiv C - C_6H_4CH_3 + 2\,CO \qquad (9.29)$$

$$\rightarrow \ (\eta\text{-}C_5H_5)(CO)_2[P(CH_3)_3]M - C \overset{\displaystyle C = O}{\underset{\displaystyle C_6H_4CH_3}{=}}$$

(M = Mo, W). Voran and Malisch (1983) reported a similar reaction for an iron-carbyne complex. Martin-Gil et al. (1979) observed a case in which the ketene ligand resulting from the interaction of a carbyne ligand with CO was bridge-bonded (μ) to two metal atoms:

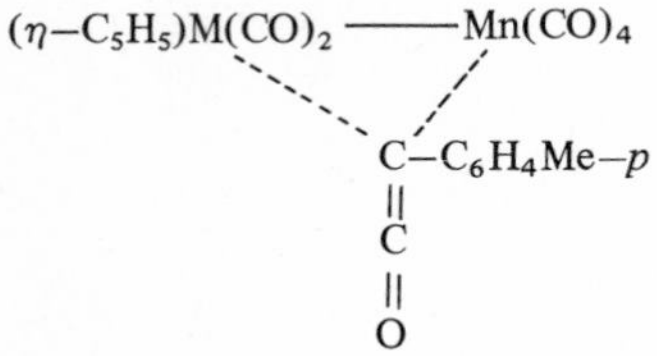

$$(\eta\text{-}C_5H_5)M(CO)_2 \text{------} Mn(CO)_4$$
$$C - C_6H_4Me - p$$
$$\parallel$$
$$C$$
$$\parallel$$
$$O$$

(M = Mn or Re).

Carbene complexes apparently tend to give η^2-ketene complexes when reacted with CO. The complex $(\eta^5\text{-}C_5H_4Me)(CO)_2$diphenylcarbenemanganese reacted with CO under high pressure (650 atm) according to

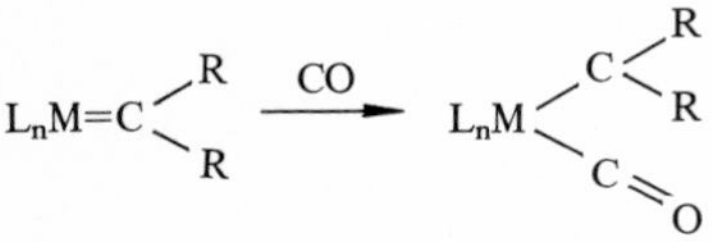

$$L_nM = C \overset{\displaystyle R}{\underset{\displaystyle R}{<}} \ \xrightarrow{\ CO\ }\ L_nM \overset{\displaystyle C \overset{R}{\underset{R}{<}}}{\underset{\displaystyle C = O}{<}} \qquad (9.30)$$

(Herrmann and Plank, 1978; Herrmann, 1982). Figure 9.12 shows the crystal structure of the interesting dihaptoketene complex. Under high pressure conditions, the η^2-ketene complex could be hydrogenolytically degraded to diphenylacetaldehyde and diphenylethanol.

Carbene ligands bonded to two metals (μ-methylene ligands) might be better models for surface carbene species. In fact, Weinberg and coworkers detected μ-methylene ligands on a Ru(001) single crystal surface when diazomethane was decomposed in high vacuum at 200 °C (George et al., 1983). Pettit and coworkers prepared the μ-methylenediiron complex:

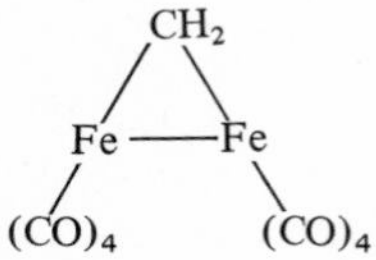

$$\begin{array}{c} CH_2 \\ \diagup \quad \diagdown \\ Fe \text{------} Fe \\ \diagup \qquad \diagdown \\ (CO)_4 \qquad (CO)_4 \end{array}$$

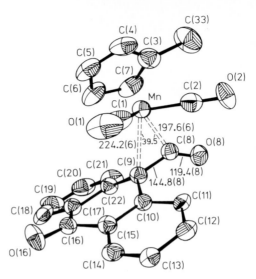

Fig. 9.12. Molecular structure of the
η^2-ketene derivative obtained according
to eq. (9.30) (Herrmann, 1982).
Reproduced by permission of Verlag
Chemie

Treated with a mixture of H_2 ($\simeq 14$ atm) and CO ($\simeq 28$ atm) at 60 °C in
solution, methane (81%) and only small amounts of acetaldehyde were formed
(Summer et al., 1980). The same complex treated with methanol (Keim et al.,
1981) resulted in methyl acetate (80%), leading the authors to the suggestion of
a μ-ketene intermediate:

It cannot be completely excluded that any of these interesting reactions take
place occasionally during a Fischer-Tropsch synthesis. But a look at Table 9.5
shows that, most probably, they do not play any dominant role. Although only
$\simeq 20\%$ of the deposited ^{13}C atoms had reacted (i.e. no exhaustion), 62% of the
C_3 hydrocarbons (in the Co case) did not contain ^{13}C, and the hydrocarbons
$> C_3$ apparently did not show any significant ^{13}C content. It appears more
probable that the small amounts of carbide based low hydrocarbons are
formed along the lines indicated in Sect. 9.5.2, rather than by the various
suggested possible chain initiation steps discussed in this Section.

9.5.5 Secondary Reactions

Whereas, at short residence time of the synthesis gas in the reactor, the major product of the FT synthesis consists of α-olefins, at longer residence time other products appear in the series

linear paraffins > β-olefins > methyl-branched paraffins

(see Fig. 9.1). A metal-hydride species, as formulated in Scheme 9.1 for the carrier of the kinetic chain, can safely be assumed to be responsible for these secondary transformations of the primary α-olefins. Presumably, the α-olefins become coordinatively bonded to such metal–H centers, and undergo hydrogenation, as well as isomerization, as known from homogeneous catalysis:

$$M-H + \alpha\text{-alkene} \rightarrow M\text{-alkyl} \underset{\xrightarrow{}\;\beta\text{-alkene} + M-H}{\overset{H_2 \, \rightarrow \, \text{alkane} + M-H}{\text{<}}} \cdot \qquad (9.31)$$

At a first sight there seems to be an enigma: Why do the metal-alkyl species occurring in the course of chain growth (cf. Scheme 9.1) resist hydrogenation and isomerization, whereas those proceeding from the coordination of α-olefins to metal-hydride species (eq. 9.31) do undergo these reactions?

The explanation resides in the preferred orientation of an α-olefin when inserted into the metal–H (or metal–R) bond. With group VIII metals, there is a remarkable tendency to the "anti-Markownikoff" mode of insertion; regioselectivities of 70% to almost 100% are the rule (Henrici-Olivé and Olivé, 1976):

$$\begin{array}{ccc} \text{H} & & \text{CH}_3 \\ | & & | \\ \text{M} + \text{CH}_2{=}\text{CHR} & \rightarrow & \text{M}-\text{CH} \\ & & | \\ & & \text{R} \end{array} \qquad \text{"anti-Markownikoff"}. \qquad (9.32)$$

The migratory CO insertion is much slower for an isoalkyl ligand than for a linear alkyl ligand (see e.g. Henrici-Olivé and Olivé, 1977). Thus, whereas in normal FT growth the CO insertion takes place before any other reaction can occur, the isoalkyl groups proceeding from the coordination of α-olefins are prone to hydrogenation and double bond migration.

Another source of alkanes cannot be ruled out, at least under certain conditions. At high ratios H_2/CO, metal-alkyl species formed in steps 8, 13 ... may undergo oxidative addition of H_2 with subsequent reductive elimination of the alkane; this would be in competition with CO insertion and further chain growth, e.g.:

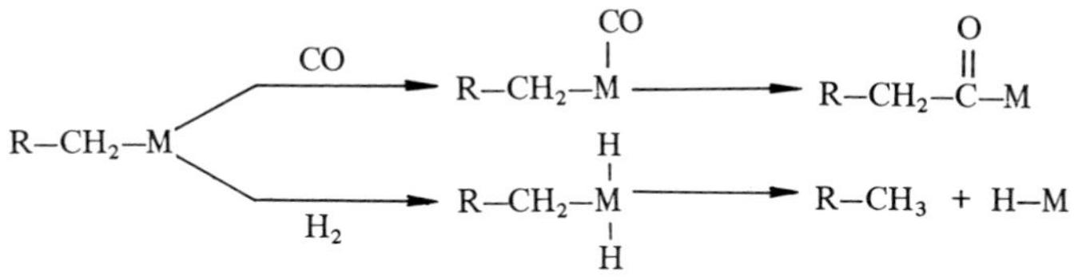

$$
R-CH_2-M \left\langle \begin{array}{l} \xrightarrow{CO} \quad R-CH_2-\overset{\overset{\displaystyle CO}{|}}{M} \longrightarrow R-CH_2-\overset{\overset{\displaystyle O}{\|}}{C}-M \\[2em] \xrightarrow[H_2]{} \quad R-CH_2-\overset{\overset{\displaystyle H}{|}}{\underset{\underset{\displaystyle H}{|}}{M}} \longrightarrow R-CH_3 \ + \ H-M \end{array} \right.
$$

Branching takes place if after the insertion of an α-olefin further chain growth takes place, e.g. for propene:

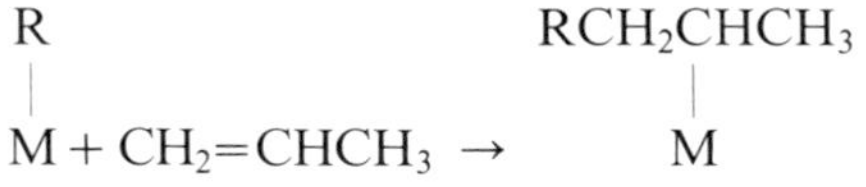

$$
\overset{\overset{\displaystyle R}{|}}{M} + CH_2{=}CHCH_3 \ \rightarrow \ \overset{\overset{\displaystyle RCH_2CHCH_3}{|}}{M}
$$

(R = H or growing chain). The corresponding reaction with ethylene gives a mere chain prolongation by two carbons; with 1-butene, an ethyl branch will result. However, the "polymerizability" of α-olefins decreases rapidly with the chain length (Henrici-Olivé and Olivé, 1977). In fact ethyl branches are rare in the FT products, and longer branches are essentially absent (see Table 9.3). Internal olefins can safely be assumed not to coordinate under the conditions of the FT synthesis.

As one may expect, the rate of consumption of CO, in moles per volume unit (cm^3) of catalyst bed and unit time (h), is highest when the residence time of the synthesis gas on the catalyst is lowest (see Table 9.6). At longer residence time, coordination of α-olefins and secondary reactions block the active centers.

Table 9.6. Influence of space velocity (residence time) on the rate of CO consumption.

Space velocity[a] (h^{-1})	Residence time (sec)	CO Conversion[a] (%)	CO Consumption ($10^3 \times mol/cm^3 \times h$)
78	46.1	84.6	0.9
337	10.7	42.1	2.0
2830	1.5	10.1	3.4

[a] Data of Pichler et al. (1967); 32% (vol.) CO; normal pressure; ca. 200 °C.

9.6 Influence of Dispersity of Metal Centers

The metallic state has been assumed frequently as essential for the catalysts of the FT synthesis, and transition metal single crystals have been used as models for catalysts with regard to bonding and structure of adsorbed carbon monoxide, hydrogen and their reaction products (see Chap. 2, 3 and 4). Much work has been invested to correlate catalytic activity (or properties believed to be associated with activity) with the peculiarities of different crystallographic planes. Steps, kinks, holes, etc., in otherwise "flat" surfaces have been found to increase activity. Evidently, metal atoms situated at such irregularities have less metal-metal coordination, and hence more sites available to coordinate reactants.

Yokohama et al. (1981) were able to show that crystallinity is not a prerequisite for catalytic activity. These authors prepared ribbons of amorphous Fe−Ni alloys, containing inclusions of phosphorus or boron atoms to shift crystallization to relatively high temperature. At temperatures well below the crystallization temperature, it was found that the activity of the amorphous state was up to several hundred times higher than that of the crystalline state, for the same composition. Again, the amorphous state, like steps or kinks on crystalline surfaces, can be expected to have a higher number of exposed atoms with an incomplete coordination sphere.

Another relevant observation is related to catalysts showing strong metal-support interaction (SMSI; see Sect. 6.2.2). TiO_2 supported catalysts in the SMSI state do not appreciably adsorb neither hydrogen nor CO (see Table 6.1) nor do they show detectable CO bands in the presence of carbon monoxide (Vannice and Wong, 1981), and yet these catalysts are an order of magnitude more active for CO hydrogenation than comparable catalysts on Al_2O_3 or SiO_2 carriers. These findings have been interpreted in terms of a small number of more or less isolated active sites (Sect. 6.2.2).

Yates et al. (1979) reported evidence for isolated mononuclear centers in the case of a 2% Rh/Al_2O_3 catalyst, although the catalytic activity of these centers was not investigated specifically (Sect. 3.3).

Convincing evidence for mononuclear species, as active species has been provided by Brenner and Hucul (1980), and has been discussed in Sect. 7.2. The coordination site requirement for a mechanism as depicted in Schemes 7.1 and 9.1, as well as evidence for the availability of up to three coordination sites at specially exposed metal centers have also been contemplated there. The reader is referred to Sect. 7.2.

Summarizing this brief discussion, evidence appears to accumulate for catalytic activity to reside on mononuclear metal centers rather than on metal surfaces. Further evidence will be provided in Sect. 9.7.2.

9.7 Influence of the Temperature and Pressure

9.7.1 Hydrocarbons versus Oxygenates

Since the early work of Fischer and his coworkers it is known that the products of the hydrogenation of carbon monoxide depend very sensibly on *temperature* and *pressure*. Based on the very instructive compilations of Storch et al. (1951),

Pichler and Krüger (1973) and Falbe (1977), as well as on the work of several other authors to be referenced individually, the following overall pictures emerge:

1. Cobalt Catalysts

At normal pressure and temperatures of 200–300 °C, linear hydrocarbons are the main product. The primary products of this synthesis are α-olefins; only at long residence times in the reactor they undergo secondary reactions (see Fig. 9.1). At medium pressures (5–30 atm, 180–250 °C), the average molecular weight increases, and the olefin content decreases in favor of alkanes. Under all conditions small amounts of alcohols are present.

2. Iron Catalysts

The general picture is similar to that observed with cobalt catalysts. Fischer and Tropsch's original hydrocarbon synthesis operated at normal pressure and 250–300 °C. The only presently operative commercial process (Sasol, South Africa) works at 24–25 atm and 220–340 °C, and produces mainly hydrocarbons (between 75 and 95%, according to process design, with 50–60% of the hydrocarbons being olefins).

However, conditions have been found, where the amount of linear primary alcohols can be greatly enhanced. One particular requirement appears to be a more moderate temperature. Thus, for a temperature range of 180–200 °C, at 10–30 atm and at relatively low space velocity ($100–300$ h^{-1}), some 40–50% of the product has been claimed to be linear alcohols. Large amounts of olefins and smaller amounts of alkanes are also present. At higher pressures (100–300 atm), and similar temperatures, but at very short residence times (space velocity $2000–10,000$ h^{-1}), even higher yields of alcohols have been reported, especially after subsequent catalytic hydrogenation of the product over a Cu/Cr catalyst, when overall yields of over 90% of alcohols are claimed (Kagan et al., 1966). Apparently, large amounts of aldehydes are formed under these high pressure conditions, which are then hydrogenated to alcohols.

3. Ruthenium Catalysts

At 1–10 atm (225–275 °C) ruthenium catalysts give hydrocarbons of the usual type, with a Schulz-Flory molecular weight distribution (Kellner and Bell, 1981), for which, however, a rather low average degree of polymerization, 2–3, can be estimated from the growth probability α, in the range of 0.5–0.7. At 30 atm and 241 °C, a hydrocarbon product with an extremely narrow molecular weight distribution has been observed (Madon, 1979), for which an average degree of polymerization of 10 can be calculated (vide infra). At 1000 atm and 132 °C, however, high molecular weight polymethylene, with molecular weight up to 100,000, has been obtained (Pichler et al., 1964); at pressures up to 2000 atm even molecular weights of the order of 10^6 have been claimed (Pichler and Krüger, 1973). Interestingly, at somewhat lower temperatures (100–120 °C), but still at high pressures (1000 atm), considerable amounts (30–40%) of oxygenated products of low molecular weight are observed, predominantly aldehydes and alcohols; evidence is reported that the alcohols are formed in a secondary reaction from the corresponding aldehydes. At 132–140 °C, however, the amount of oxygenated products is negligible (Bellstedt, 1971).

None of the various mechanisms suggested thus far in the literature for the Fischer-Tropsch synthesis has taken into account and/or sufficiently interpreted these facts. It appears, however, that the CO insertion mechanism on single, mononuclear surface metal complexes, as suggested in Scheme 9.1, can accommodate these apparently enigmatic features of the synthesis. In this Chapter we want to point particularly to those assumed intermediates and individual steps of the complex overall mechanism, which may be responsible for the observed pressure and temperature effects.

Fischer-Tropsch catalysts are generally prepared from transition metal oxides, which are reduced with hydrogen to the metallic state, up to 96%. Under the conditions of the synthesis, an active surface metal center may then be expected to be surrounded by the following ligands: other metal atoms, perhaps oxygen from the original oxide or from the carrier, the growing chain, carbon monoxide and/or hydrogen, and possibly carbidic carbon. In the following we shall summarize all ligands, except the growing chain, by L_x. Note that L_x comprises CO and/or hydrogen; the two ligands may be expected to displace each other under the reaction conditions (see Eq. 7.3).

A growing alkyl chain has, in principle, the following reaction possibilities:

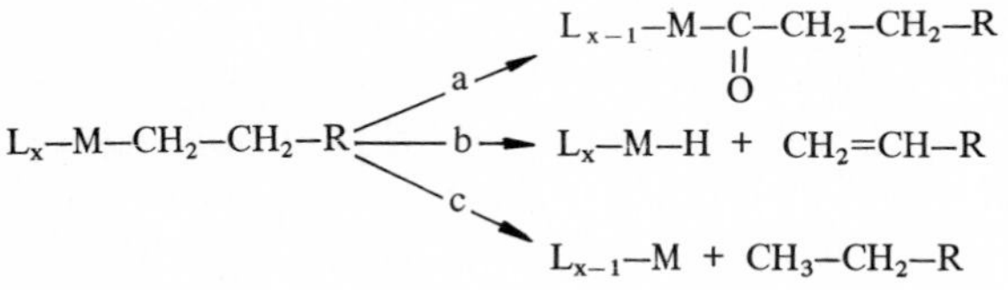

$$
L_x{-}M{-}CH_2{-}CH_2{-}R
\begin{cases}
a & L_{x-1}{-}M{-}\underset{\underset{O}{\|}}{C}{-}CH_2{-}CH_2{-}R \\
b & L_x{-}M{-}H \;+\; CH_2{=}CH{-}R \\
c & L_{x-1}{-}M \;+\; CH_3{-}CH_2{-}R
\end{cases}
\tag{9.33}
$$

Reaction a (CO insertion) enlarges the chain by one carbon unit. From homogeneous coordination chemistry it is well known that this reaction does not require a free coordination site, since the alkyl group migrates to a CO coordinated to the same metal (Sect. 5.3.1).

Reaction b (β-H abstraction) produces an α-olefin and a metal hydride. From homogeneous coordination and metallorganic chemistry it is known that this reaction does require a free coordination site, in cis position to the alkyl ligand (Sect. 5.4.1). Wilkinson and coworkers have shown that a metal-alkyl bond can be stabilized, and β-H abstraction virtually suppressed, by preventing free sites at the metal center (Thomas et al., 1968).

Reaction c (hydrogenolysis of the metal-alkyl bond) does not require a free site, it does, however, require that hydrogen ligands are present in the coordination sphere of the metal center; its result is an alkane. As already mentioned, α-olefins are the primary products at low pressure, whereas alkanes are produced by hydrogenation in a secondary reaction. Reaction *c* does not appear to play an important role under these conditions, presumably because of competition from reactions *a* and *b*.

Apparently, the availability of free coordination sites at low pressure permits the β-H abstraction, as well as hydrogenation of the primary olefins, to take place. With increasing pressure, reaction *b* may be expected to be increasingly inhibited, since CO and/or hydrogen will maintain the coordination sites occupied. At very high pressure, the β-H abstraction can safely be assumed not to take place at all.

The hydrogenation of the acyl ligand formed in the reaction *a*, to a hydroxyalkyl ligand, was suggested to proceed in two steps, with an intermediate aldehyde coordinatively bonded to the metal center (Scheme 9.1):

$$L_x-M-\underset{\underset{O}{\|}}{C}-R \xrightarrow{d} L_{x-1}-M\cdots\underset{\underset{O}{\|}}{\overset{CH-R}{\|}} \xrightarrow{e} L_{x-2}-M-\underset{\underset{OH}{|}}{CH}-R. \tag{9.34}$$

At low pressure, the aldehyde does not leave the active center (no aldehydes have been found among the products at low pressure by Pichler et al., 1967). At high pressures, CO may well be able to displace occasionally an aldehyde molecule from the catalytic center, making then reaction *d* into an alternative molecular weight determining step. Aldehydes have, in fact, been found among the products at high pressure (Kagan et al., 1966; Bellstedt, 1977).

The hydroxyalkyl ligand, if formed in reaction *e*, has also several possible reaction paths, one of which molecular weight determining:

$$
\begin{array}{l}
\quad\quad\quad L_{x-1}-M-\underset{\underset{O}{\|}}{C}-\underset{\underset{OH}{|}}{CH}-R \\
\;\;\overset{f}{\nearrow} \\
L_x-M-\underset{\underset{OH}{|}}{CH}-R \xleftarrow{\;\;} \overset{g}{\longrightarrow} L_{x-1}-M{=}CH-R + H_2O \\
\quad\quad\quad\quad \searrow h \quad\quad\quad\quad \llcorner\!\!\longrightarrow L_{x-2}-M-CH_2-R \\
\quad\quad\quad L_{x-1}-M + HOCH_2-R
\end{array} \tag{9.35}
$$

Reaction *f* (CO insertion into a metal-hydroxyalkyl bond) has never been detected in heterogeneous processes; it is, however, well known in homogeneous medium, where it leads to hydroxyaldehydes, polyalcohols, etc. (see Chap. 10).

Reaction *g* is assumed to be the dehydration step in Scheme 9.1, leading to an alkyl ligand via a carbene intermediate, and hence crucial for chain growth.

Reaction *h* (hydrogenolysis of the metal-hydroxyalkyl bond) leads to an alcohol which then leaves the center.

None of these three reactions requires a free coordination site. Hence, pressure should not have a marked influence on the fractions of hydroxyalkyl ligands that proceed either to chain growth (g) or to elimination as an alcohol (h). In fact alcohol syntheses have been claimed in the intermediate pressure range (10–30 atm, Falbe, 1977, Chap. 10.1), as well as at high pressures (100–300 atm, Kagan et al., 1966). However, the finding that lower temperatures favor the formation of alcohols (all other conditions comparable) appears to indicate a considerable difference in activation energy between g and h ($E_g > E_h$). At lower temperatures, both reactions are supposedly relatively slow (as indicated by a low overall reaction rate), the ratio k_h/k_g determining the average molecular weight of the alcohols. With increasing temperature, the rate of reaction g increases much more rapidly than that of h, to the extent that at the usual temperatures for the synthesis of hydrocarbons (250–300 °C), the formation of alcohols according to h does not play any significant role (Falbe, 1977, Chap. 8). It may be noteworthy that a large fraction of alcohols among the products has been reported only for iron, but not for cobalt catalysts, and for ruthenium catalysts only as the consequence of a secondary hydrogenation of aldehydes. Apparently, the ratio k_h/k_g is quite different for the three metals (Fe > Co > Ru).

9.7.2 The Particular Case of Ruthenium Catalysts

The discussion carried on thus far would take care of most of the phenomena observed with Fe and Co catalysts. But what makes Ru so different? Let us start with the high pressure case reported by Pichler et al. (1964).

The experiments were carried out in an autoclave, at 1000 atm and 132 °C; the Ru catalyst was dispersed in nonane; the result was polymethylene. The authors reported that the polymer was of dark color and contained ruthenium which could not be removed, neither by filtration, nor by reprecipitation. After fractionation, all fractions were dark and contained ruthenium. IR spectroscopic analysis indicated that this polymer was present as

$$(\text{polymer chain}) - \text{Ru(CO)}_x. \tag{9.36}$$

Only with hydrogen at 140–150 °C and 150–200 atm, or by boiling with 0.2 n KOH in isopropanol, could the ruthenium be separated from the polymer. As stated by the authors, the reaction conditions (pressure and temperature) were close to those favorable for the formation of volatile ruthenium carbonyl complexes. The very high molecular weight shows that β-H abstraction (reaction b) was essentially absent, as expected at the high pressure used. Evidently, the chain growth was never interrupted, and "living polymers" were present.

The appearance of aldehydes at this high pressure (1000 atm), but at somewhat lower temperature (100–120 °C), as reported by Bellstedt (1971) appears plausible (see Eq. 9.34 and accompanying text). At temperatures > 132 °C, no more oxygenated products are found. This seems to indicate that

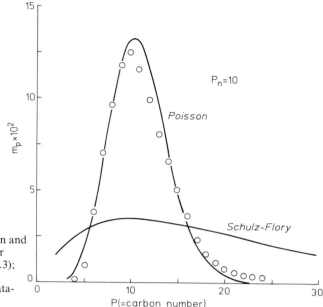

Fig. 9.13. Theoretical Poisson and Schulz-Flory distributions for $P_n = 10$ (solid lines, cf. Fig. 9.3); experimental data by Madon (1979), obtained with a Ru catalyst at 30 atm, 241 °C

at this temperature the rate of reaction e (Eq. 9.34) is sufficiently high, so that it can take place before the aldehyde is able to leave the complex.

The general picture of "living polymers" on ruthenium catalysts finds an interesting confirmation by data of Madon (1979). Working at 30 atm and 241 °C with a Ru/Al_2O_3 catalyst, Madon obtained hydrocarbons, the molecular weight distribution of which could not be described in the usual way by the Schulz-Flory equation, but was considerably narrower. Flory (1940) has shown that a polymerization with no termination gives a product ("living polymer") approaching a molecular weight distribution described by a Poisson equation, as shown in Section 9.3.1 (Eq. 9.16). Figure 9.13 shows an amazingly close correspondence of the experimental data published by Madon and the theoretical Poisson distribution typical for "living polymers", with an average degree of polymerization of $P_n = 10$.

These findings, together with Pichler's observation of alkylruthenium carbonyls (Formula 9.36) represent an additional strong evidence that the Fischer-Tropsch chain growth takes place on single, mononuclear metal centers (stable carbonyl complexes). Consequently, multiple-coordinated species such as those suggested by Pichler (Eq. 9.25), as well as similar species proposed by Muetterties and Stein (1979) and Deluzarche et al. (1982), might not be significant for hydrocarbon growth in the FT synthesis.

The fact that ruthenium catalysts do not necessarily form volatile carbonyl complexes under all conditions (Ru losses through volatile complexes can be prevented by exhaustive reduction of a Ru catalyst prior to its use, as reported by Bellstedt (1971), does not invalidate the above discussion concerning molecular growth centers.

The results of Kellner and Bell (1981) who reported hydrocarbon products fitting a Schulz-Flory distribution with Ru catalysts at 1 atm and 240 °C, striking as it may appear at first sight in view of Pichler's and Madon's data, fit smoothly into the general picture given here, since at 1 atm a free coordination site on the active metal center can be assumed to be available. Hence, the β-H abstraction can take place, controlling the molecular weight.

9.8 The Role of Alkali Promoters

The addition of small amounts of alkali metal compounds, in particular K_2O or K_2CO_3, to an iron catalyst for the Fischer-Tropsch synthesis is known to produce substantial changes in the catalytic performance. The synthesis activity as well as the average molecular weight of the hydrocarbons formed are increased; moreover, the olefinic fraction in the hydrocarbon effluent is improved (Storch et al., 1951; Dry, 1969).

An interpretation of these phenomena, based on the electron-donor capacity of K_2O, has been forwarded by Dry et al. (1969, 1972). Electrons donated to the metal are assumed to enhance the adsorption of CO, while reducing that of H_2; weakening of the CO bond would facilitate the attack of hydrogen. Reduced hydrogenation ability, on the other hand, would lead to higher molecular weight (whereby the chain terminating step is assumed to be hydrogenolysis of an alkyl-metal bond), as well as to a higher olefinic fraction. Similar suggestions have been made by Rähse and Schneidt (1972) and Kagan et al. (1966).

This interpretation is, however, not quite compatible with the views of molecular catalysis as outlined in Scheme 9.1, Sect. 9.5.3. Transition metal carbonyl complexes are bases, and tend to form complexes with Lewis acids (electron acceptors) rather than with electron donors (Hieber, 1970; Shriver, 1975). For a number of transition metal carbonylates, as well as donor (e.g. alkyl) substituted neutral carbonyl complexes, it has been shown by X-ray structural analysis, IR, and other measurements that the interaction with an electron acceptor (Mg^{2+}, K^+, Na^+, Li^+, AlX_3, etc.) occurs via the electron-rich oxygen of the carbonyl group (Henrici-Olivé and Olivé, 1982b, and references therein). The various X-ray and IR studies have demonstrated that such interaction strengthens the metal-carbon bond, and weakens the carbon-oxygen bond of the involved carbonyl ligand.

Collman et al. (1972, 1978) have shown that the close interaction of an alkali cation with a carbonyl ligand of the alkyliron carbonyl $Na^+[RFe(CO)_4]^-$ greatly favors the migratory insertion of the CO ligand into the metal alkyl bond:

$$Na \cdots \overset{\displaystyle \overset{\textstyle Na^+}{\underset{\textstyle |}{\overset{\textstyle O^-}{:}}}}{OC{-}Fe(CO)_3R} \;\rightarrow\; R{-}\overset{\displaystyle \overset{\textstyle Na^+}{\underset{\textstyle |}{\overset{\textstyle O^-}{:}}}}{C}{-}Fe(CO)_3. \tag{9.37}$$

These authors found that in THF at 25 °C, the cations Li^+ or Na^+ cause the insertion reaction to occur 2 to 3 orders of magnitude faster than the rate observed if the cation was trapped in a crown ether, or if the bulky cation $[(C_6H_5)_3P]_2N^+$ was used instead. In the resulting acyl complex, the cation is associated with the oxygen of the acyl group. Collman et al. assume that the rate of alkyl migration is so dramatically increased by the presence of electron acceptors Li^+ or Na^+, because the latter stabilize the coordinatively unsaturated intermediate.

The presence of the acceptor cation may help to dissipate temporarily electron density released onto the metal through the loss of one π-acceptor ligand (CO), in the course of the formation of the acyl ligand (cf. Berke and Hoffmann, 1978, who indicated that π-acceptor ligands on the metal in the migratory plane would have a similar effect). It should, however, also be taken into account that the interaction of an alkali cation with the oxygen of a carbonyl group leads to a lowering of the energy of the lowest unoccupied molecular orbital (frontier orbital), of the carbonyl group, as shown by Anh (1976). This certainly contributes to facilitating the insertion reaction.

Other pertinent observations have been made by Shriver and coworkers (Correa et al., 1980) with neutral alkylmanganese and alkyliron carbonyl complexes, where the migratory insertion was greatly accelerated by the addition of the Lewis acid $AlBr_3$. Again, the accepting aluminum is associated with the acyl oxygen in the resulting acyl complex. In the absence of gaseous CO the $AlBr_3$ serves simultaneously as an electron donor to the transition metal atom, a bromine anion filling the coordination site vacated in the course of the insertion reaction, whereby a five-membered ring structure is formed:

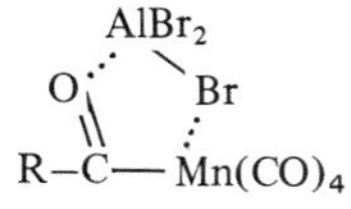

Remarkably, Shriver and coworkers found that a comparable acceleration of the migratory CO insertion takes place if a pentane solution of $Mn(CH_3)(CO)_5$ or $(C_5H_5)Fe(CH_3)(CO)_2$ is injected onto a solid surface of dehydroxylated alumina.

A further example of promotion of the insertion of CO into a metal-carbon bond has been provided by Lindner and von Au (1980). The rhenacycle compound shown in Eq. (9.38) undergoes ring expansion by CO insertion in the presence of $AlBr_3$:

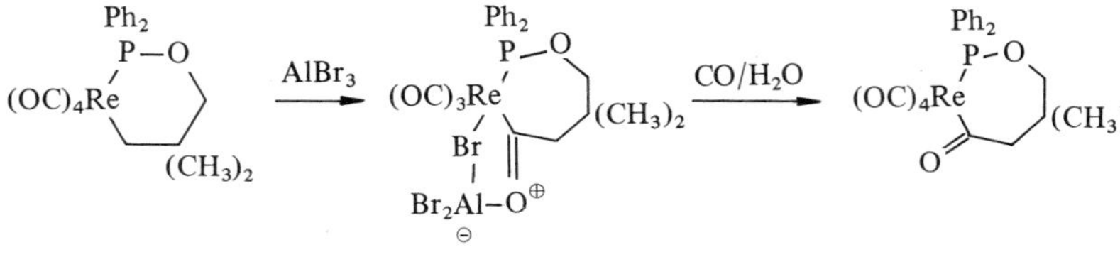

After hydrolysis and elimination of the Lewis acid, the rhenacycle II was isolated. In the absence of AlBr$_3$ the insertion does not take place even at 800 atm and 100 °C. The strong interaction between Al and the carbonyl in (I) (Eq. 9.38) can be judged from the value of the CO stretching frequency which is 1393 cm^{-1} for I and 1608 cm^{-1} for II.

With all this excellent literature work in mind, the present authors suggested that the effect of alkali promoters in the Fischer-Tropsch synthesis might be similar to that in the above mentioned carbonyl complexes (Henrici-Olivé and Olivé, 1982b). One has to consider that under the conditions of the FT synthesis (presence of CO, H$_2$, CO$_2$, H$_2$O, high temperature) the surface iron atoms are in an environment different from that in the original (reduced) catalyst. In particular, there will be surface iron species featuring one or several CO ligands. The potassium promoter, on the other hand, in whatever form it might have been added to the catalyst, is certainly present as K$^+$, even if it would have been transformed to metallic potassium in the reductive activation of the catalyst. Thus, if a surface alkyliron carbonyl species has a suitable positioned neighboring K$^+$ ion, the migratory insertion reaction should be greatly accelerated. Along the same lines, the further reaction of the acyl group (still in interaction with the cation) with hydrogen should also be promoted, due to the lowering of the energy of the CO frontier orbital.

Assuming that the β-H transfer from the growing chain to the transition metal, as the growth ending step, is not affected by the alkali promoter, an increase of the average molecular weight follows cogently from the increased growth rate, since the average degree of polymerization (i.e. the number of C atoms in the hydrocarbon chain) is given by the ratio of the rate of chain propagation to the rate of chain transfer

$$P_n = \frac{r_p}{r_{tr}} = \frac{d[CH_2]/dt}{d[olefin]/dt} .$$

The amount of alkali metal ions in the promoted catalysts is low; generally less than 0.01 g K is added per g Fe (Dry et al., 1969). Hence, surface iron carbonyl complexes that have no suitable neighboring K$^+$ ion should always be present, and these should produce chain growth and concomitant chain lengths according to unpromoted catalysts. In other words: a bimodal molecular weight distribution may result from promoted catalysts. Relevant molecular weight distribution data are scarce; however, there is some evidence for bimodal distribution produced by alkali-promoted iron catalysts as compared to a normal distribution produced by unpromoted cobalt catalysts (Anderson, 1956). Clear examples of bimodal molecular weight distributions have been reported for the reaction products obtained with alkali-promoted cobalt catalysts (Madon and Taylor, 1979).

Finally, the observed increase of the olefin fraction of the effluent fits also into the picture. With the reasonable assumption that a metal hydride is responsible for the hydrogenation of the olefins, and that the latter must be coordinated to the metal center prior to reaction, the following trivial explanation can be visualized: The reasons given above for the acceleration of the insertion of CO into the metal-alkyl bond should apply equally well to the chain initiation (insertion of CO into a metal-hydrogen bond):

$$
(L_x)M\!-\!H
\begin{cases}
\overset{CO}{\longrightarrow} \quad \overset{\displaystyle CO}{\underset{\vdots}{(L_x)\!\overset{}{M}\!-\!H}} \longrightarrow \text{Chain initiation}\\[2em]
\underset{R-C=C}{\longrightarrow} \quad \overset{\displaystyle R-C=C}{\underset{\vdots}{(L_x)\!\overset{}{M}\!-\!H}} \longrightarrow \text{Hydrogenation}
\end{cases}
$$

(L_x = all other ligands).

Hence, in the case of the promoted catalyst, chain initiation can compete more favorably with hydrogenation than in the non-promoted catalyst. In conclusion the remarkable effect of potassium promoters in the Fischer-Tropsch synthesis may satisfactorily be interpreted based on well documented phenomena, and fitting smoothly within the framework of molecular catalysis.

9.9 Product Selectivity

9.9.1 Consequences of Schulz-Flory Molecular Weight Distribution

The mechanism of the FT synthesis, as a step-by-step polymerization with chain growth and chain transfer, imposes severe restrictions to the selectivity with regard to the molecular weight of the product. The validity of the Schulz-Flory distribution function (cf. Sect. 9.3) determines the selectivity for a given chain length as

$$
m_P = (\ln^2 \alpha)\, P\, \alpha^P \tag{9.39}
$$

with m_P = mass fraction of product having degree of polymerization P, and α = probability of chain growth. Figure 9.14 shows this dependence for a few selected values of P. The Fig. indicates that C_2 peaks at $\alpha \simeq 0.4$, C_4 at $\alpha \simeq 0.6$, C_6 at $\alpha \simeq 0.7$, etc. However, whichever the value of α, hydrocarbons of other degrees of polymerization are simultaneously present in amounts determined by Eq. (9.39). A typical Schulz-Flory distribution was represented in Fig. 9.3, for the particular case of $\alpha = 0.905$, resulting in an average degree of polymerization of $P_n = 10$.

At high CO conversion (long residence time) this relatively simple relationship tends to be complicated by the products of secondary reactions, in

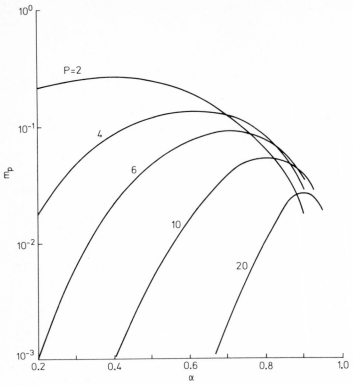

Fig. 9.14. Consequences of a Schulz-Flory molecular weight distribution: molar fraction of selected chain lengths as a function of the growth probability α, calculated according to eq. (9.39)

particular those leading to branched molecules. As discussed in Sect. 9.1, certain catalysts (Fe > Co) and process design (fluid-bed > fixed-bed) have a higher propensity to give branched material. For the cases of such branched products, Friedel and Anderson (1950) introduced a correction factor with an adjustable parameter into their molecular weight distribution function (cf. Sect. 9.3.1); the resulting equation proved to be very useful in organizing a large amount of experimental data (see e.g. Storch et al., 1951).

To a certain degree it is possible to vary the parameter α, and hence the molecular weight, by process variables. Thus, alkali promoters shift the chain length to higher values (Sect. 9.8). The ratio H_2/CO in the gas feed has also some influence (the higher this ratio the lower the average molecular weight, cf. Sect. 9.1). The effect of the ratio metal/carrier has been mentioned in Sect. 9.5.3: the lower this ratio the smaller the average molecular weight. This principle has been used by Chauvin and coworkers to shift the molecular weight distribution to the low side (Commereuc et al., 1980). Values of α in the range of 0.2–0.4 can be determined from their data, obtained with noble metal (Rh, Ir, Ru, Os) catalysts on alumina or silica. Slightly higher values

Table 9.7. Synthesis of ethylene on Al_2O_3 supported catalysts. $CO/H_2 = 2$; space velocity 2500 h^{-1}; p = 140 torr (GP, 1968).

Metal	T (°C)	CO Conversion (%)	CH$_4$ (mol %)	C$_2$H$_4$ (mol %)
Co	350	14.4	8.9	91.1
Ni	300	16.9	20.9	79.1
Pt	300	17.5	35.1	64.9

(0.5–0.6) can be estimated from the somewhat scattered data with several iron catalysts.

In a system where the β-hydrogen abstraction rate is extremely high compared with the rate of chain growth, one might expect that the former takes place as soon as a β-H becomes available, i.e. when a CH$_3$CH$_2$ ligand is formed (cf. Scheme 9.1, step 13). This situation is well known in homogeneous catalysis, in the case of the selective dimerization of olefins (see e.g. Henrici-Olivé and Olivé, 1977). In the FT synthesis this would lead preferentially to ethylene, apart from methane that has several formation routes (e.g. the carbide route, Sect. 9.5.2; decomposition of low olefins, Sect. 9.2.2). One such case has actually been claimed for a reaction at subatmospheric pressure (140 torr; $\leqq$ 1% of metal on Al_2O_3 and/or SiO_2), and at a ratio CO/H_2 of 2 (GP, 1968). Methane and ethylene are the only hydrocarbons produced and the selectivity for ethylene is amazingly high (see Table 9.7).

There are reports in the literature indicating that the average molecular weight can also be influenced somewhat by the particle size of the metal (dispersion of the metal on the carrier). It was found that α decreases with the particle size (Everson et al., 1980; see Table 9.8), and that this is accompanied by a decrease in specific activity. The Schulz-Flory distribution remains valid (Kellner and Bell, 1982).

The influence of the particle size on the specific activity is a somewhat controversial issue. Increase (Vannice, 1975, 1976), as well as decrease (vide supra), with diminishing particle size have been claimed, and a variety of interpretations have been forwarded. Takasu et al. (1978) have cautioned that

Table 9.8. Probability of chain growth, α, as a function of metal dispersion. Catalyst: Ru on γ-Al_2O_3; T = 250 °C; H_2/CO = 2; p = 12 atm; Conversion $\simeq$ 22% (Everson et al., 1980).

Dispersion (%)	Crystallite size (nm)	α
78	1.3	0.65
70	1.5	0.66
47	2.2	0.67
16	6.4	0.73

in some cases the preparation of the samples (impregnation of, or precipitation onto, support) may have made it difficult to obtain clean metal surfaces and to characterize them correctly. These authors used a model Pd catalyst prepared by evaporating metal onto a carbon film. The mean diameter of the metal particles was controlled by the amount of metal deposited, and determined by electron microscopy. The electron energy levels within the metal were characterized by X-ray photoelectron spectroscopy. The bonding energy of the $3d_{5/2}$ electrons was found to increase, from the bulk metal value, as the particle size decreases, the major change (1.5 eV) taking place in the range from 1.8 to 1.0 nm diameter. The energy of the valence band in the Pd particles is assumed to shift with that of the $3d_{5/2}$ band (Takasu et al., 1978). This is suggested as the cause of the experimental finding that the activation energy of a catalytic test reaction (H_2/D_2 exchange) increases with decreasing particle size. These interesting findings appear to indicate that the observed variations of activity with particle size are due to activity changes of the individual metal centers, rather than to a decrease in the number of active sites, as other authors had suggested (e.g. Topsøe et al., 1981; Kellner and Bell, 1982). Alterations in the electronic structure of small metal particles that are attributable to strong metal-support interactions have, in fact, repeatedly been described (e.g. Tauster et al., 1981; Aral et al., 1982, see also the discussion of strong metal-support interaction in Sect. 6.2.2). It appears plausible that such changes translate into variations of the chain growth probability (cf. Eq. 9.5, Sect. 9.3.1):

$$\alpha = r_p/(r_p + r_{tr}).$$ (9.40)

However, it should be taken into account that the changes in specific activity (r_p per metal center) and in α cannot be expected to be proportional, since r_p and r_{tr} are most probably influenced to a different degree and possibly in opposite direction (see Sect. 5.5). In fact, a decrease of α by 10% has been reported to be accompanied by a decrease in specific activity of about an order of magnitude (Kellner and Bell, 1982).

Finally, it may be noted that a certain increase of α can be obtained by the addition of ethylene to the feed gas (Henrici-Olivé and Olivé, 1982, and references therein). This procedure may be considered as copolymerization of two "monomers", ($CO+H_2$) and $CH_2=CH_2$. Adapting Eq. (9.40) to this situation one obtains:

$$\alpha = \frac{\Sigma\, r_p}{\Sigma\, r_p + r_{tr}}$$ (9.41)

where $\Sigma\, r_p$ is the sum of the rates of the different growth steps. (Compare the similar situation in Sect. 9.5.2, Eq. 9.22.)

9.9.2 Deviations from the Schulz-Flory Distribution

Selectivity is a crucial problem in FT synthesis. Not always is the whole broadness of products, as dictated by the Schulz-Flory equation, really wanted. Thus, for gasoline, a branched product in the range C_5-C_{11} would be desirable; for heating purposes, a high content of low molecular weight hydrocarbons, particularly methane, ethane and propane, is necessary; for the production of plasticizers, detergents, synthetic lubricants, etc., linear α-olefins in the range C_6-C_{16} would be the most favorable. Thus, one of the most recent branches of Fischer-Tropsch chemistry consists of designing ways to circumvent the Schulz-Flory distribution law, in order to improve selectivity.

One such case has been shown in Sect. 9.7.2: the narrow Poisson distribution found for Ru-grown "living polymers". A promising approach appears to be the trapping of catalytic transition metal species in the cavities of zeolites. A brief introduction into the possibilities of molecular shape selectivity has been given in Sect. 6.2.3 (see also Fig. 6.3 for an example of zeolite structure). The molecular weight limiting properties of zeolites in the synthesis of hydrocarbons has been exploited and developed by Mobil in their methanol based process to gasoline (Voltz and Wise, 1977; see also Sect. 11.3).

The first application of zeolites, as a chain limiting catalyst, in the FT synthesis was reported by Nijs et al. (1979). A RuNa zeolite was prepared by partial exchange of the Na^+ ions (cf. Eq. 6.9) in a faujasite type zeolite with $Ru(NH_3)_6CL_3$. The ruthenium was then reduced with hydrogen at 300 °C. The FT synthesis was run under conditions where a usual Ru/SiO_2 catalyst would give a product with 60% > C_{12}, and a Schulz-Flory distribution. The zeolite catalyst generated a product with less than 1% > C_{12}. In the case of a Ru/La zeolite, a chain limitation at C_5 was observed. Ballivet-Tkatchenko and Tkatchenko (1981) prepared zeolite catalysts by thermally decomposing metal carbonyls (of Fe, Co, Ru) within the cavities of zeolites; they also observed selective formation of hydrocarbons in the C_1-C_9 range. A Schulz-Flory plot shows the usual picture up to C_9, and thereafter a sharp drop, indicating that higher hydrocarbons are only formed in traces.

Fraenkel and Gates (1980) cautioned that the high temperature reduction of the transition metal species within the zeolite cavities might lead to a partial destruction of the original aluminosilicate structure. These authors used cadmium vapor for the reduction of Co^{2+} ions exchanged into different types of zeolites. The intactness of the zeolite framework structures was checked by X-ray diffraction. On a zeolite characterized by relatively small cavities, exclusively propylene was formed (151 °C; $\simeq$ 6 atm; $CO:H_2 = 1$). Cd species remaining in the zeolite cavities, either alone or combined with Co species, are assumed to participate in the catalytic action. On a faujasite type zeolite, the main product was n-butane, the C_4-C_7 mixture constituting $\simeq$ 70% of the hydrocarbon product and the distribution deviating considerably from the Schulz-Flory equation.

The size limiting effect of the zeolite cavities may be visualized as the consequence of a steric hindrance of the migratory insertion of CO into the growing chain. As the cavity gets "crowded", this step may become increasingly difficult, its probability eventually dropping to zero. The abstraction of a β-H will then free the molecule from its catalytic "birth place", the metal center.

Other reasons for deviations from Schulz-Flory statistics have been reviewed by Jacobs and van Wouwe (1982): deposition of high molecular weight hydrocarbons on the catalyst bed, operation in transient conditions, artifacts in sampling, secondary reactions, and others.

9.10 Conclusion

The revival of the Fischer-Tropsch synthesis, in the aftermath of the oil embargo in the early seventies, has brought about a great deal of interesting chemistry. However, considerable research and development work, in particular with regard to selectivity for certain products of the synthesis, needs still to be done in order to have this coal-based route to hydrocarbons available when the need reappears.

9.11 References

Anderson RB (1956) Catalysts for the Fischer-Tropsch Synthesis; in: Emmett PH (ed) Catalysis, Vol. 4, Hydrocarbon Synthesis, Hydrogenation and Cyclization. Reinhold, New York (cf. Figs 76, p 359 and 27, p 110)
Anderson RB, Emmett PH (1956) Catalysis. Reinhold, New York, Vol 4
Anh NT (1976) Tetrahedron Lett, 155
Aral M, Ishikawa T, Nishiyama Y (1982) J Phys Chem 86:577
Azam KA, Frew AA, Lloyd BR, Manojlovic-Muir L, Muir KW, Puddephatt RJ (1982) J Chem Soc Chem Commun 614
Ballivet-Tkatchenko D, Tkatchenko I (1981) J Mol Catal 13:1
van Barneveld WAA, Ponec V (1978) J Catal 51:426
Bellstedt F (1971) PhD Thesis-Karlsruhe, West Germany
Berke H, Hoffmann R (1978) J Amer Chem Soc 100:7224
Biloen P, Helle HN, Sachtler WMH (1979) J Catal 58:95
Biloen P, Helle HN, van den Berg FGA, Sachtler WMH (1983) J Catal 81:450
Bradley JS (1979) J Amer Chem Soc 101:7417
Brady RC, Pettit R (1980) J Amer Chem Soc 102:6182
Brady RC, Pettit R (1980) J Amer Chem Soc 103:1287
Brenner A, Hucul DA (1980) J Amer Chem Soc 102:2484
Brötz W, Rottig W (1952) Z Elektrochem 56:896
Casey CP (1979) CHEMTECH 378, and references therein
Chisholm MH, Clark HC, Johns WS, Ward JEH, Yasufuku K (1975) Inorg Chem 14:900
Collman JP, Cawse JN, Bramman JI (1972) J Amer Chem Soc 94:9505
Collman JP, Finke RG, Cawse JN, Bramman JI (1978) J Amer Chem Soc 100:4766
Commereuc D, Chauvin Y, Hugues F, Basset JM, Oliver D (1980) J Chem Soc Chem Commun 154
Correa F, Nakamura R, Stimson RE, Burwell RL, Shriver DF (1980) J Amer Chem Soc 102:5112
Dautzenberg FM, Helle HN, van Santen RA, Verbeek H (1977) J Catal 50:8
DBP (1949) 904, 891 (Ruhrchemie)

DBP (1961) 974, 811 (Ruhrchemie)

Deluzarche A, Hindermann JP, Kieffer R, Cressely J, Stupfler R, Kiennemann A (1982) Spectra 2000 10:27; Bull Soc Chim France II 329

Driessen JM, Poels EK, Hindermann JP, Ponec V (1983) J Catal 82:26

Dry ME (1969) Brennstoff-Chem 50:193

Dry ME, Hoogendoorn JC (1981) Catal Rev-Sci Eng 23:265

Dry ME, Shingles T, Boshoff LJ, Oosthuizen GJ (1969) J Catal 15:190

Dry ME, Shingles T, Boshoff LJ (1972) J Catal 25:99

Everson RC, Smith KJ, Woodburn ET (1980) Proc Natl Meet S Afr Inst Chem Eng 3rd 2 D-1

Falbe J (ed) (1977) Chemierohstoffe aus Kohle. Georg Thieme Verlag, Stuttgart

Falbe J, Frohning CD (1982) J Mol Catal 17:117

Fischer F, Pichler H (1937/51) Abh Kenntn Kohle 13:407

Fischer F, Pichler H (1939) Brennstoff-Chem 20:41, 221

Fischer F, Tropsch H (1923) Brennstoff-Chem 4:276

Fischer F, Tropsch H (1924) Brennstoff-Chem 5: 201, 217

Fischer F, Tropsch H (1926) Brennstoff-Chem 7:97

Flory PJ (1936) J Amer Chem Soc 58:1877

Flory PJ (1940) J Amer Chem Soc 62:1561

Fraenkel D, Gates BC (1980) J Amer Chem Soc 102:2478

Friedel RA, Anderson RB (1950) J Amer Chem Soc 72:1212, 2307

George PM, Avery NR, Weinberg WH, Tebbe FN (1983) J Amer Chem Soc 105:1393

GP (1968) Auslegeschrift 1271,098

Henrici-Olivé G, Olivé S (1974) Adv Polymer Sci 15:1

Henrici-Olivé G, Olivé S (1976a) Angew Chem 88: 144; Angew Chem Int Ed Engl 15:136

Henrici-Olivé G, Olivé S (1976b) Topics in Current Chemistry 67:107

Henrici-Olivé G, Olivé S (1977) Coordination and Catalysis. Verlag Chemie, Weinheim, New York

Henrici-Olivé G, Olivé S (1978) J Mol Catal 4:379

Henrici-Olivé G, Olivé S (1979) US Patent 4,179,462 (to Monsanto Company)

Henrici-Olivé G, Olivé S (1982a) J Mol Catal 16:111

Henrici-Olivé G, Olivé S (1982b) J Mol Catal 16:187

Henrici-Olivé G, Olivé S (1983) J Mol Catal 18:367

Herington EFG (1946) Chem Ind 347

Herrmann WA (1982) Angew Chem 94:118; Angew Chem Int Ed Engl 21:117

Herrmann WA, Plank J (1978) Angew Chem 90:557; Angew Chem Int Ed Engl 17:525

Hieber W (1970) Adv Organomet Chem 8:7

Jacobs P, van Wouwe D (1982) J Mol Catal 17:145

Kagan YuB, Bashkirov AN, Morozov LA, Kryukov YuB, Orlova NA (1966) Neftekhimiya 6:262

Keim W, Röper M, Strutz H (1981) J Organometal Chem 219:C 5

Kellner CS, Bell AT (1981) J Catal 70:418

Kellner CS, Bell AT (1982) J Catal 75:251

Klabunde U, Fischer EO (1967) J Amer Chem Soc 89:7142

Koch H, Hillerath H (1941) Brennstoff-Chem 22:135, 145

Kreissl FR, Uedelhoven W, Eberl K (1978) Angew Chem 90:908; Angew Chem Int Ed Engl 17:859

Kummer JT, DeWitt TW, Emmett PH (1948) J Amer Chem Soc 70:3632

Lee BS (1982) Syn Fuels from Coal. AIChE Monograph Series No 14, Vol 78

Lindner E, von Au G (1980) Angew Chem 92:843; Angew Chem Int Ed Engl 19:824

Madon RJ (1979) J Catal 57:183

Madon RJ, Taylor WF (1979) Effect of Sulfur on the Fischer-Tropsch Synthesis; Alkali-Promoted Precipitated Cobalt-Based Catalysts. In: Kugler EL, Steffgen FW (eds) Hydrocarbon Synthesis from Carbon Monoxide and Hydrogen. Adv Chem Ser 178

Markó L, Bakos J (1974) J Organometal Chem 81:411

Martin-Gil J, Howard JAK, Navarro R, Stone FGA (1979) J Chem Soc Chem Commun 1168

Mazzi U, Orio AA, Nicolini M, Marzotto A (1970) Atti Accad Peloritana Pericolanti, Cl Sci Fis Mat Nat (Engl) 50:95

Merta R, Ponec V (1969) J Catal 17:79

Muetterties EL, Stein J (1979) Chem Rev 79:479

Niemantsverdriet JW, van der Kraan AM, van Dijk WL, van der Baan HS (1980) J Phys Chem 84:3363

Nijs HH, Jacobs PA (1980) J Catal 65:328; 66:401

Nijs HH, Jacobs PA, Uytterhoven JB (1979) J Chem Soc Chem Commun 180, 1095

Nyholm RS (1965) in: Sachtler WMH, Schuitt GCA, Zwietering P Proc. 3rd Internat. Congr. Catal. Vol. 1, p. 25. North Holland, Amsterdam

Patard G (1924) Compt red Acad Sci 179:1330

Pichler H, Buffleb H (1940) Brennstoff-Chem 21:257, 273, 285

Pichler H, Krüger G (1973) Herstellung flüssiger Kraftstoffe aus Kohle, Gersbach & Sohn, München

Pichler H, Schulz H (1970) Chem Ing Techn 42:1162

Pichler H, Firnhaber B, Kioussis D, Dawalla A (1964) Makromol Chem 70:12

Pichler H, Schulz H, Elstner M (1967) Brennstoff-Chem 84:1967

Pichler H, Schulz H, Kühne D (1968) Brennstoff-Chem 49:344

Rähse W, Schneidt D (1972) Ber Bunsenges Phys Chem 77:727

Raupp GB, Delgass WN (1979) J Catal 58:348, 361

Roelen O (1948) Angew Chem 60:524

Roelen O (1978) Erdöl und Kohe 31:524

Rofer-de-Poorter C (1981) Chem Rev 81:447

Sabatier P, Senderens JB (1902) Compt rend Acad Sci 134:514, 689

Sachtler JWA, Koo JM, Ponec V (1979) J Catal 56:284

Schäfer-Stahl H (1980) Angew Chem 92:761; Angew Chem Int Ed Engl 19:729

Schlesinger MD, Benson HE, Murphy E, Storch HH (1954) Ind Eng Chem 46:1322

Schulz GV (1935) Z phys Chem 29:299; 30:375

Schulz GV (1936) Z phys Chem 32:27

Schulz H, Rao BR, Elstner M (1970) Erdöl Kohle 23:651, and literature therein

Shriver DF (1975) J Organometal Chem 94:259

Soviet Patent (1973) 386,899; 386,900

Storch HH, Golumbic N, Anderson RB (1951) The Fischer-Tropsch and Related Syntheses. Wiley, New York

Summer CE, Riley PE, Davis RE, Pettit R (1980) J Amer Chem Soc 102:1752

Takasu Y, Unwin R, Tesche B, Bradshaw AM, Grunze M (1978) Surface Sci 77:219

Takasu Y, Akimaru T, Kasahara K, Matsuda Y, Miura H, Toyoshima I (1982) J Amer Chem Soc 104:5249

Takeuchi A, Katzer JR (1981) J Phys Chem 85:937

Tauster SJ, Fung SC, Baker RTK, Horsley JA (1981) Science 211:1121

Thomas K, Osborn JA, Powell AR, Wilkinson G (1968) J Chem Soc A 1801

Topsoe H, Topsoe N, Bohlbro H, Dumesik JA (1981) Stud Surf Sci Catal 247

Tropsch H, Koch H (1929) Brennstoff-Chem 10:337

Ullmann's Enzyklopädie der Technischen Chemie (1957 a, b) Urban und Schwarzenberg Verlag, München, Berlin. a) Vol. 9; b) Vol. 3

US Patent (1973) 3,726,926 (W.R. Grace and Co.)

Vannice MA (1975) J Catal 37:449, 462; 40:129

Vannice MA (1976) J Catal 44:152

Vannice MA, Wang SY (1981) J Phys Chem 85:2543

Voltz SE, Wise JJ (eds) (1977) Development Studies on Conversion of Methanol and Related Oxygenates to Gasoline. Orn 1/FE-1, Springfield NTIS

Voran S, Malisch W (1983) Angew Chem 95:151; Angew Chem Int Ed Engl 22:151

Weiss K, Fischer EO (1976) Chem Ber 109:1868

Wender I, Friedman S, Steiner WA, Anderson RB (1958) Chem Ind 1694

Yang CH, Massoth FE, Obkad AG (1979) Adv Chem Series 178:35

Yates JT, Duncan TM, Worley SD, Vaughan RW (1979) J Chem Phys 70:1219

Yokohama A, Komiyama H, Inoue H, Matsumoto T, Kimura HM (1981) J Catal 68:355

Ziegler K, Holzkamp E, Breil H, Martin H (1955) Angew Chem 67:541

10 Homogeneous CO Hydrogenation

Methanation (Chap. 7), methanol formation (Chap. 8), and Fischer-Tropsch synthesis (Chap. 9) are essentially domains of heterogeneous catalysis, although methanol is also produced, together with methyl formate and/or ethylene glycol with the homogeneous $HCo(CO)_4$ system at high pressure, as described in Sect. 8.1.

In this Chapter those reactions will be discussed which are confined to the homogeneous (liquid) medium: hydroformylation of olefins and of formaldehyde, the catalytic synthesis of glycol and higher polyalcohols from CO and H_2, and related reactions.

10.1 Hydroformylation of Olefins (Oxo Reaction)

The hydroformylation of olefins detected by Roelen (1938) is one of the most important commercial examples of homogeneous catalysis with transition metal compounds. Under carbon monoxide and hydrogen pressure (mostly $CO:H_2 = 1:1$) of over 100 atm, at $150-180\,°C$, the olefin is transformed to an aldehyde having one carbon atom more than the starting olefin, according to the overall equation:

$$RCH=CH_2 + CO + H_2 \xrightarrow{\text{Cat}} RCH_2CH_2CHO. \tag{10.1}$$

This process is by now a textbook item (e.g. Henrici-Olivé and Olivé, 1977; Collman and Hegedus, 1980; Cotton and Wilkinson, 1980), and is covered by many excellent reviews (Paulik, 1972; Orchin and Rupilius, 1972; Markó, 1974; Pino, 1980) to which the reader is referred for a broad coverage of this interesting reaction. In the present context, limitation to some mechanistic details and a comparison with the systems treated thus far appears appropriate.

Oxides and complexes of several transition metals, in particular those of cobalt and rhodium, have been applied as catalyst precursors. Rhodium catalysts are more active, permitting operation at lower pressure and temperature, even at room temperature and 1 atm, in the case of phosphine modified

Rh catalysts (Brown and Wilkinson, 1970). The soluble, active species formed under the reaction conditions may be formulated generally as $HM(CO)_mL_n$, with M = transition metal ion, L = neutral ligand, such as phosphine, phosphite, or amine, $m \geq 1$, $n \geq 0$, and $(m + n) = 3$ or 4. Its formation, as well as its stability in solution, require a certain CO partial pressure (Markó, 1961; Sisak et al., 1983). The necessary CO pressure is the higher, the higher the reaction temperature. In the case of $HCo(CO)_4$, for instance, at least 50 atm are required in the usual temperature range; otherwise metallic cobalt would precipitate. Wilkinson's phosphine modified Rh catalyst is the exception, since it works at room temperature and no excessive CO pressure is necessary.

Scheme 10.1 shows, in a somewhat simplified manner, the generally agreed course of the hydroformylation of a terminal olefin. (For the sake of clarity, $(CO)_m$ and L_n ligands are omitted; it is understood that the insertions of olefin and of CO are preceded by the coordination of these molecules to the metal center.

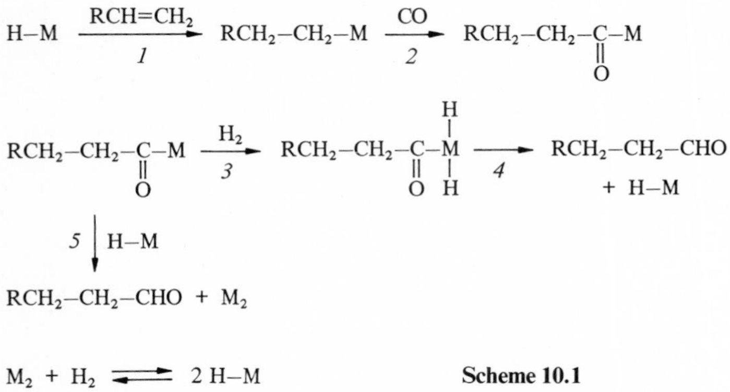

$$H-M \xrightarrow[\;1\;]{RCH=CH_2} RCH_2-CH_2-M \xrightarrow[\;2\;]{CO} RCH_2-CH_2-\underset{\underset{O}{\|}}{C}-M$$

$$RCH_2-CH_2-\underset{\underset{O}{\|}}{C}-M \xrightarrow[\;3\;]{H_2} RCH_2-CH_2-\underset{\underset{O}{\|}\;\underset{H}{|}}{C}-\overset{\overset{H}{|}}{M} \xrightarrow{\;4\;} RCH_2-CH_2-CHO + H-M$$

$$5 \downarrow H-M$$

$$RCH_2-CH_2-CHO + M_2$$

$$M_2 + H_2 \rightleftharpoons 2\,H-M \qquad\qquad \textbf{Scheme 10.1}$$

Two possible routes have been discussed in the literature for the last step, the transformation of the acyl-metal species to the aldehyde: hydrogenolysis after oxidative addition of a H_2 molecule (steps 3 and 4), or reaction with a second catalyst species (step 5). The former route was preferred by Heck and Breslow (1961), who first suggested the mechanism outlined in Scheme 10.1. Step 5, involving two mononuclear species, has been considered by Breslow and Heck (1960), but was discarded on the ground of the low concentration of that species. However, spectroscopic evidence for the importance of step 5 has been reported since by Penninger and coworkers (Bowen et al., 1975) for the particular case where $H-M = HCo(CO)_4$. The authors investigated the hydroformylation of octene, at $p_{H_2} \simeq 50$ atm; $p_{CO} \simeq 10$ atm, $T \simeq 100\,°C$, with $Co_2(CO)_8$ as the catalyst precursor. In situ IR spectroscopy in a high pressure cell was used as the tool. First, it was found that the declustering of $Co_2(CO)_8$ with H_2 is a slow reaction under the reaction conditions:

$$Co_2(CO)_8 \xrightarrow{H_2} 2\,HCo(CO)_4. \qquad\qquad (10.2)$$

The establishment of the equilibrium required several hours. Most of the cobalt (84%) was then present as $HCo(CO)_4$. Addition of octene-1 caused a sudden decrease of the IR peaks characteristic for $HCo(CO)_4$, whereas those of $Co_2(CO)_8$ increased. This concentration change gradually reverted as the conversion of octene to nonal proceeded. These observations were convincingly interpreted assuming partial (or even exclusive) aldehyde formation according to step 5 (Scheme 10.1), with a relatively slow subsequent delivery of $HCo(CO)_4$ from equilibrium (10.2), at least under the prevailing reaction conditions. A further evidence for the availability of step 5 (Scheme 10.1) comes from the stoichiometric reaction of $HCo(CO)_4$ with olefins, resulting in the corresponding aldehyde (Orchin and Rupilius, 1972), which takes place at $0\,°C$ (1 atm N_2):

$$RCH=CH_2 + 2\,HCo(CO)_4 \;\rightarrow\; RCH(CHO)CH_3 + Co_2(CO)_7. \qquad (10.3)$$

In this case the hydridocobaltcarbonyl is the only source of CO as well as of hydrogen.

Martin and Baird (1983) tried to decide between steps $(3 + 4)$ and 5 (Scheme 10.1), in the particular case of phosphine modified Co complexes. They prepared separately the acetyl complex $CH_3C(O)Co(CO)_3(PMePh_2)$ and the hydrido complex $HCo(CO)_3(PMePh_2)$. Their conclusion was that in this particular case step 5, (the intermolecular reductive elimination) did not play any important role. But, unfortunately, their results are not quite conclusive. If the two complexes were brought together at room temperature and normal pressure, conversion of the hydrido complex to the dimer

$$2\,HCo(CO)_3L \;\rightarrow\; H_2 + Co_2(CO)_6L_2$$

"was the faster process". This is not amazing, since mononuclear hydrido-cobaltcarbonyl complexes require a certain CO pressure to be stable at room temperature. (See, however, Eq. (10.3) which refers to $0\,°C$.) Hydrogen at normal pressure did not react with the acyl complex, but did so at 35 atm (no CO pressure). Since the acetyl complex is pentacoordinated, it requires decarbonylation before oxidative addition of hydrogen can take place:

$$CH_3\underset{\substack{\| \\ O}}{C}Co(CO)_3L + H_2 \xrightarrow{\;-CO\;} CH_3\underset{\substack{\| \;\; | \\ O\;\; H}}{C}Co(CO)_2\overset{H}{\underset{|}{L}}.$$

It is doubtful whether this reaction would take place easily under a considerable CO pressure, as under actual hydroformylation conditions. It is also undecided whether the hydrido complex would have reacted with the acyl complex under pressure.

A comparison between Scheme 10.1 for the hydroformylation with Scheme 9.1 for the Fischer-Tropsch synthesis (Sect. 9.5.3), shows that the two types of reactions are closely related. Evidently, the insertion of a CO molecule into a metal-carbon bond with the formation of an acyl-metal species is a common step in both reactions (step 2 in Scheme 10.1; steps 9, 14, ... in Scheme 9.1). It may appear enigmatic that under closely-related reaction conditions and with similar catalysts, the further course of the reactions is thus different. In hydroformylation, the aldehyde is formed and immediately leaves the complex; no further chain growth takes place. In the Fischer-Tropsch reaction, on the other hand, the aldehyde (assumed as a plausible intermediate) does not leave the coordination sphere of the metal, but subsequent reaction with the remaining hydrogen ligand (step 4, Scheme 9.1) opens the way to further chain growth via steps 5, 7, 8, etc. The chain length is controlled by the ratio of the rates of "chain propagation" (CO insertion) and "chain transfer" reactions (steps 6, 12, 15, ...), leading to a Schulz-Flory distribution of molecuar weights, as shown in Chap. 9.

Based on the findings of Penninger and his coworkers, the present authors have suggested that the decisive difference between the homogeneous and the heterogeneous process is precisely the availability of free, mobile, very reactive hydrido-metal species in solution, which makes step 5 (Scheme 10.1) the only important mode of reaction of the acyl-metal species in the homogeneous system (Henrici-Olivé and Olivé, 1977/1978). The simultaneous formation of the dinuclear metal complex, M_2 [$Co_2(CO)_8$ in the case of $HCo(CO)_4$ as active species], makes further reaction in the coordination sphere impossible, and causes the aldehyde to leave the metal center. In the heterogeneous system, on the other hand, the $M-H$ species are fixed at their surface sites, and cannot encounter any acyl-metal species, also fixed at the surface. Thus, the oxidative addition of molecular hydrogen to the latter is the only means of reaction.

The relative importance of the two routes (reactions 3 + 4 and reaction 5) will be discussed further in Chap. 12.

It may be interesting to consider a report by Moffat (1970), which at first sight appears to contradict the present suggestion: high yields of aldehydes have been obtained on a heterogeneous cobalt catalyst (cobalt carbonyl on a polymeric carrier), under the conditions of the OXO reaction. It was, however mentioned that some of the cobalt had leached into the liquid phase as $HCo(CO)_4$ and $Co_2(CO)_8$; hence the hydroformylation conditions as required by the suggested mechanism (availability of $HCo(CO)_4$ for the reduction of the acyl species) are established. The phenomenon is not surprising taking into account that most hydroformylation catalysts on the basis of cobalt are formed from insoluble cobalt compounds under the influence of a CO/H_2 atmosphere and at temperatures above 50 °C (Falbe, 1967).

Vapor phase hydroformylation of ethylene and propylene over cobalt and rhodium clusters deposited on ZnO or MgO support has also been reported (Ichikawa, 1979). Gas mixtures of CO, H_2 and olefin (1:1:1) were circulated over the catalyst bed, at normal pressure and 100−180 °C. In this case it may be safely assumed that the high concentration of olefin in the gas phase caused the displacement of the aldehyde from the active center, so that Fischer-Tropsch growth could not take place. This view is corroborated by up to 50% hydrogenation of the olefin parallel to the hydroformylation.

Present and future work in the field of hydroformylation of olefins, as far as showing in the literature, is mainly concerned with methods of improving the ratio of linear to branched aldehydes among the products, while reducing the hydrogenation of the olefin to alkane to a minimum (see e.g. Clark and Davies, 1981; Saus et al., 1983), or with stabilizing the active catalyst so that product distillation at elevated temperature would not destroy it (e.g. Matsumoto and Tamura, 1983).

10.2 Hydroformylation of Formaldehyde

The reaction of formaldehyde, coordinatively bonded to a metal center, with CO and hydrogen is, so to speak, the connecting link between all variants of catalytic hydrogenation of CO.

The stoichiometric hydroformylation of formaldehyde with $HCo(CO)_4$, at $0\,°C$ in CH_2Cl_2 and 1 atm of CO, has been achieved by Roth and Orchin (1979); see Sect. 5.3.9. In this case, as in the case of the stoichiometric hydroformylation of an olefin (Eq. 10.2), the metal hydride was the only source of hydrogen. Glycol aldehyde was the predominant (up to 90%) product, indicating that of the two possible modes of insertion of CH_2O into the metal$-$H bond (Eqs. 5.31 and 5.32) only one (5.31) was operative.

Spencer (1980) investigated the mechanism of the catalytic formaldehyde hydroformylation with rhodium catalysts of the type $RhCl(CO)L_2$ (L = triaryl-phosphine), at $110\,°C$ and $80-130$ atm, $CO/H_2 = 1$. Although the mechanism was found, in general, similar to that of the olefin hydroformylation, a solvent effect was observed, which does not occur in the latter, and may be linked to the polar character of the substrate formaldehyde. The major product of the hydroformylation of CH_2O was glycol aldehyde, but the hydrogenation product, CH_3OH, was also present. The ratio hydroformylation/hydrogenation was strongly solvent dependent. The formation of glycol aldehyde was favored only in N,N-disubstituted amides as solvent, where up to 50% of the CH_2O was transformed to glycol aldehyde, whereas in solvents such as acetone, benzene or ethanol only hydrogenation was observed. N,N-disubstituted amides such as dimethylformamide or dimethylacetamide, may be considered as electron donating ligands if present in the coordination sphere of a metal center. Nevertheless, there was no ligand influence on the insertion mode of the formaldehyde molecule in the sense indicated above in Eq. (8.3). Spencer found only traces of methyl formate, indicating that the insertion mode was not governed by a donor ligand. The mechanism suggested by him is represented in Scheme 10.2. (Spencer proposed the transformation VII → II as

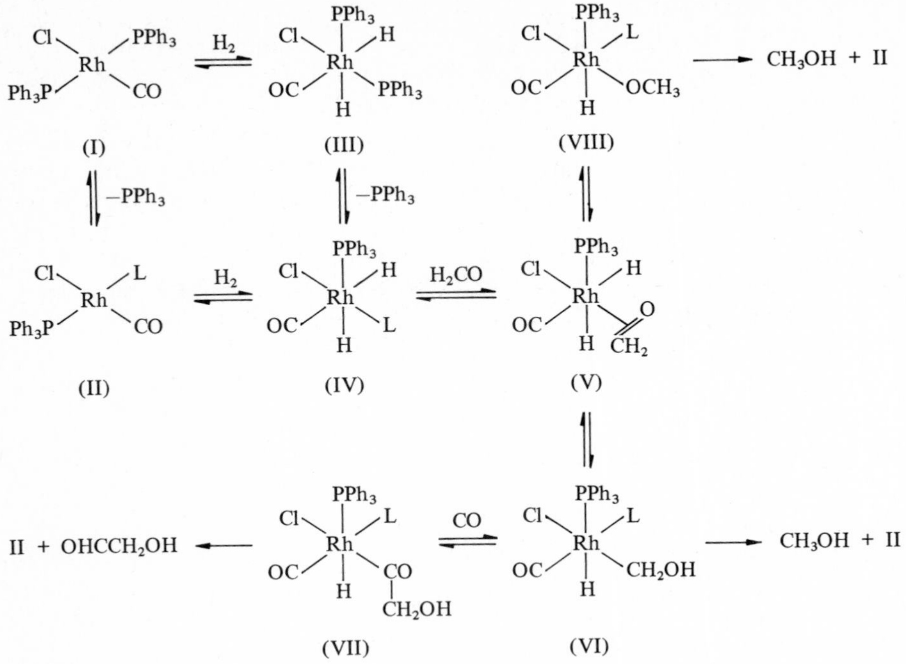

Scheme 10.2

The oxidative addition and reaction sequence (I → III → VIII → CH$_3$OH + II; I → II; III → IV → V → VI → VII → II + OHCCH$_2$OH):

- (I): Cl, PPh$_3$ / Rh / Ph$_3$P, CO
- (III): Cl, H / Rh(PPh$_3$) / OC, PPh$_3$, H
- (VIII): Cl, L / Rh(PPh$_3$) / OC, OCH$_3$, H → CH$_3$OH + II
- (II): Cl, L / Rh / Ph$_3$P, CO
- (IV): Cl, H / Rh(PPh$_3$) / OC, L, H
- (V): Cl, H / Rh(PPh$_3$) / OC, H, O, CH$_2$
- (VII): Cl, L / Rh(PPh$_3$) / OC, CO, H, CH$_2$OH → II + OHCCH$_2$OH
- (VI): Cl, L / Rh(PPh$_3$) / OC, CH$_2$OH, H → CH$_3$OH + II

Equilibria indicated: I ⇌ III (H$_2$); III → VIII (PPh$_3$); I ⇌ II (−PPh$_3$); III ⇌ IV (−PPh$_3$); II ⇌ IV (H$_2$); IV ⇌ V (H$_2$CO); V → VI; VI ⇌ VII (CO)

L = amide

an equilibrium, because excess glycol aldehyde reduced the yield; blocking of coordination sites by the bidentate ligand OHCCH$_2$OH appears more probable as cause for this inhibition, therefore this step was not formulated as an equilibrium in Scheme 10.2.) The oxidative addition of H$_2$ to the square planar catalyst complex II is assumed by Spencer to precede formaldehyde coordination. Thus, CH$_2$O has to displace the solvent ligand L in an octahedral complex, i.e. no orienting influence by L can be expected. The influence of the donor solvent ligand may then become operative in the step VI → VII, since migratory insertion reactions are generally favored by donor ligands (Sect. 5.5.1). The HOCH$_2$CHO/CH$_3$OH yield ratio should be determined by the relative rates of CO insertion and reductive elimination of CH$_3$OH. The availability of the route to CH$_3$OH via species VIII is left open to question; since no methyl formate was present, it may not play any role.

An *alternative mechanism* starting with a rhodium hydridocarbonyl species and involving reaction of the hydroxyacyl ligand with a second Rh−H species would be closer to present knowledge with regard to olefin hydroformylation, cf. Scheme 10.1. The transformation of Rh(CO)(Cl)(PR$_3$)$_2$ catalyst precursors to HRh(CO)(PR$_3$)$_2$ is reasonably well supported in olefin hydroformylation systems (see e.g. Henrici-Olivé and Olivé, 1977; Chap. 10.2.2). The principle of the solvent effect as discussed above (via an enhancement of the migratory CO insertion) would, of course, remain valid.

This view is corroborated by data of Chan et al. (1983), who investigated closely related systems. Starting with $HRh(CO)(PR_3)_3$ or $RhCl(CO)(PR_3)_2$, they found similar results; moreover, they could further improve the selectivity to glycol aldehyde and increase the reaction rate, by adding a strong electron donor, triethylamine ($Et_3N/Rh = 4$) to the system. (It should be mentioned, though, that the authors suggest a different mechanism, via deprotonation of the hydridorhodium complex by the amine, reaction of the anionic rhodium species with CH_2O and subsequent reprotonation to form the hydroxymethyl rhodium intermediate.)

10.3 Polyalcohols from CO + H₂, and Related Homogeneous Syntheses

Homogeneous systems have failed thus far to produce any significant amount of Fischer-Tropsch chain growth. Instead, *cobalt*, *rhodium* and *ruthenium* catalysts produce mainly methanol with varying amounts of glycol aldehyde, glycol and methyl formate under conditions of medium-high pressure (see Sect. 8.2).

The absence of Fischer-Tropsch chain growth appears to be an important prerequisite for *selectivity*, and hopes have been running high to detect selective routes to valuable oxygenates. Ethylene glycol, presently produced in a multistep process from oil, is a particularly attractive target.

Early patents (USP, 1950; USP, 1953) indicated that organic products including glycol and glycerin, could be obtained from CO + H₂ by homogeneous catalysis with *cobalt* catalysts under extreme pressure (1500−5000 atm). Subsequently, Pruett and his coworkers at Union Carbide discovered improved selectivity to glycol with soluble *rhodium* catalysts at elevated pressure (Pruett, 1977). Small amounts of glycerin were also present; the major byproduct was methanol. In over a dozen of patents, gradual improvement of the synthesis was reported (for a review see Falbe, 1980). The presence of small amounts of a "promoter" such as 2-hydroxypyridine, o-phenanthroline, or pyridine is favorable. (Presumably, these compounds act as donor ligands.) The preferred reaction conditions evolving from this work appear to be: 1500−2000 atm, 190−240 °C, tetraglyme or sulfolane as solvent, Rh concentration 0.05 − 0.3 wt%. Under such conditions a selectivity of 60−65% to polyols has been claimed.

The *rhodium* may be added in any convenient form, e.g. dicarbonylacetylacetonaterhodium (I), hexarhodium hexadecacarbonyl, or tetrarhodium dodecacarbonyl. Pruett suggested that a rhodium carbonyl cluster $[Rh_{12}(CO)_{34}]^{2-}$ which formed in situ under the reaction conditions, was the actual catalyst. In fact, such highly clustered species have been experimentally observed under high pressure. However, present opinion appears to favor declustering to mononuclear species, with or without formation of hydride, as the decisive step towards formation of a catalytically active species, at least in the case of *cobalt* and *ruthenium* catalysts (see e.g. Feder and Rathke, 1980; Fahey, 1981; Dombek, 1980).

Ruthenium catalysts are considerably less active for the polyol synthesis; under comparable reaction conditions they produce mainly methanol. But Dombek (1981) has found that the addition of an iodide (in particular KI) enhanced simultaneously the overall activity of the system and the selectivity to two-carbon products, especially ethylene glycol and ethanol. Under the most favorable conditions (sulfolane solvent, 850 atm, 180 °C) a molar ratio of $(CH_2OH)_2/CH_3OH = 1.1$ was found which shows that 2.2 times more CO was transformed to glycol than to methanol. Ethanol was identified as a secondary product, on the basis of studies of product distribution as a function of reaction time (methanol homologation, see Sect. 11.2). Small amounts of methane, glycerin, and glycol acetals of acetaldehyde and glycol aldehyde were also present. Reaction solutions after catalysis contained most of the ruthenium as $[HRu_3(CO)_{11}]^-$ or $[Ru(CO)_3I_3]^-$. Each of these complexes alone exhibits only low catalytic activity, but synthetic mixtures of the two are more active. This may indicate that a hydride as well as an iodide ligand are necessary in the active species.

The use of mixed Ru/Rh catalysts dispersed in a low-melting quaternary phosphonium salt has resulted in mixtures of glycol, glycol methyl ester, methanol and ethanol (Knifton, 1983a); however, the yield of glycol, or its derivatives was not improved, and the reaction rate was low.

Another activation of Ru catalysts was found by Dombek (1980) and Knifton (1982, 1983b): the reaction rate of the CO hydrogenation was improved if the synthesis was carried out in acetic acid as a solvent. The pressure could be lowered to 340−430 atm, at 220−230 °C. The reaction products in this case were the esters of methanol (mainly), glycol, ethanol, and glycerin (traces). The ethanol (for the formation of the corresponding ester) is assumed to be either derived from the solvent (Dombek) or formed by homologation of methanol (Knifton).

Keim et al. (1980), as well as Kiennemann et al. (1982), compared the activity of several group 8 metal catalysts under comparable conditions, at very high pressure (2000−3000 atm). In polar solvents, the series of activities is $Rh > Ru \gg Co \simeq Ir > Pt \gg Pd, Os, Fe, Ni$. In non-polar solvents, Co was found more active than Rh, Ru (Keim et al.). In most cases, selectivity is poor, a large number of oxygenates being formed. In the case of Ru in tetraglyme, at extremely high pressure (> 3000 atm), the formation of small amounts of $C_2 - C_9$ alcohols (besides the major product methanol) was reported (Kiennemann et al.). These alcohols might have been formed by normal Fischer-Tropsch growth (cf. Scheme 9.1), suggesting that part of the ruthenium has precipitated in the course of the reaction. Kiennemann et al. suggested homologation instead. This appears less probable in view of the poor homologation ability of ruthenium (see Sect. 11.2); no case of homologation of methanol to alcohols up to C_9 is reported. However, the influence of the high pressures is not quite known.

The reaction scheme leading to glycol and the other oxygenates mentioned in this Section, has been outlined in its essentials in Scheme 8.1 (Sect. 8.2). It has to be completed by the explicit inclusion of all possible further reactions of glycol aldehyde, the assumed intermediate on the way from species 5

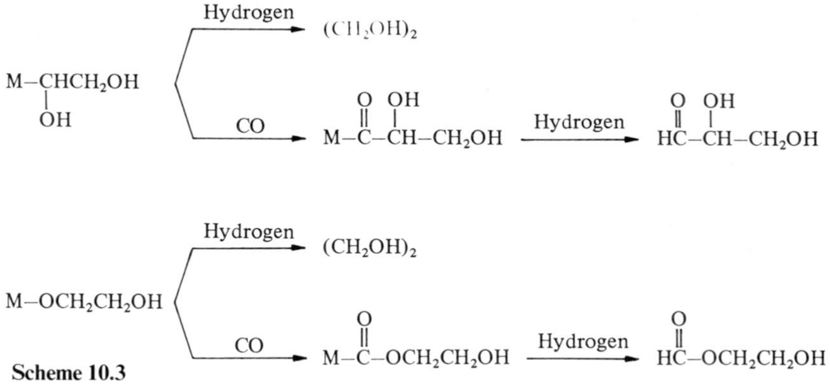

Hydrogen
(CH$_2$OH)$_2$

M–CHCH$_2$OH
   |
   OH

CO

$$M-\overset{O}{\overset{\|}{C}}-\overset{OH}{\overset{|}{CH}}-CH_2OH \xrightarrow{\text{Hydrogen}} HC-\overset{OH}{\overset{|}{CH}}-CH_2OH$$
(with $\overset{O}{\overset{\|}{HC}}$ at the right product)

Hydrogen
(CH$_2$OH)$_2$

M–OCH$_2$CH$_2$OH

CO

$$M-\overset{O}{\overset{\|}{C}}-OCH_2CH_2OH \xrightarrow{\text{Hydrogen}} \overset{O}{\overset{\|}{HC}}-OCH_2CH_2OH$$

Scheme 10.3

(Scheme 8.1) to glycol. The aldehyde may react without leaving the complex (as assumed by Parker et al., 1982; see Sect. 8.2), or after leaving and recoordination, as suggested by Fahey (1982). (It should be noted that for heterogeneous CO hydrogenation, the former way is more probable, since aldehydes have not been found among the primary products; in homogeneous systems the latter appears more plausible, see Chap. 12.)

Scheme 10.3 gives the expansion of Scheme 8.1, starting with the two species corresponding to intermediates *3* and *4* in Scheme 8.1, which result from the two insertion modes of glycol aldehyde into the metal−H bond. (For reasons of clarity, all other ligands are omitted.) The products are glycol, glycerol aldehyde and glycol monoformate. Further reaction of glycerol aldehyde along the same lines would evidently lead to glycerin. The formates (last lines in Schemes 8.1 and 10.3) apparently do not react further (no corresponding reaction products are reported).

In Scheme 10.3, all reductive elimination steps have been intentionally formulated in a vague manner, involving "hydrogen" in some form. In view of the situation in hydroformylation (Sect. 10.1 and 10.2), where it appears highly probable that the reductive elimination is an intermolecular process involving a metal hydride, it is tempting to assume that all (or most) reductive eliminations in homogeneous medium under pressure are of the intermolecular type. (This would then, of course, also apply to Scheme 8.1 which would require some alterations.) This fundamental question will be discussed in a broader context in Chap. 12.

In contrast to the heterogeneous methanation (Scheme 7.1, Sect. 7.2) and Fischer-Tropsch synthesis (Scheme 9.1, Sect. 9.5.3) where water is evolved, all homogeneous reactions leading to oxygenates (Schemes 8.1 and 10.3) take place without loss of water. Evidently, this is one of the striking differences between heterogeneous and homogeneous systems. Possible causes will also be discussed in Chapter 12. Another remarkable phenomenon is the requirement of high pressure in the case of homogeneous synthesis of methanol (Sect. 8.2) and of very high pressure for the polyalcohols (this Section). A look at the thermodynamics of these reactions makes this understandable (Table 10.1). Whereas the formation of methane, with evolution of water, is thermo-

Table 10.1. Thermodynamic data of some CO hydrogenation reactions (227 °C, 1 atm), according to Pruett (1977).

Reaction	ΔG (kJ/mol)	log K_p
$CO + 3H_2 \rightarrow CH_4 + H_2O$	− 96.1	10.06
$CO + H_2 \rightarrow CH_2O$	+ 50.6	− 5.29
$CO + 2H_2 \rightarrow CH_3OH$	+ 21.3	− 2.22
$2CO + 3H_2 \rightarrow HOCH_2CH_2OH$	+ 66.0	− 6.89
$3CO + 4H_2 \rightarrow HOCH_2CH(OH)CH_2OH$	(+ 10.9)[a]	−
$2CO + 3H_2 + 2HOAc \rightarrow AcOCH_2CH_2OAc + 2H_2O$	− 38.0[b]	− 2.0[b]

[a] estimated;
[b] Knifton (1982)

dynamically favorable, the reactions without loss of water are unfavorable, and the more so, the more carbon-carbon bonds are built. An increasingly large, unfavorable entropy term renders ΔG positive for the temperatures usually employed in these processes. Pressure is, therefore, required to shift the equilibrium. Since increasingly more moles $(CO + H_2)$ are needed to make up one mole of product, very high pressure can make up for the unfortunately high positive ΔG for the glycol and the higher polyols.

The formation of formaldehyde is also quite unfavorable at normal pressure (cf. Sect. 8.2). In this context it appears noteworthy that the catalytic formation of oxygenates by hydroformylation of formaldehyde, where this energetically costly step has been eliminated, can be carried out at 100 atm (Sect. 10.2), whereas those homogeneous syntheses starting with CO and H_2 require 300−1500 atm.

The last entry in Table 10.1 shows why the route to glycol esters, as applied by Dombek and Knifton and discussed earlier in this Section, can be operated at a relatively low pressure.

The relative selectivity to any of the oxygenates to be found in Schemes 8.1 and 10.3 depends on numerous variables, several of which have been mentioned in Sect. 8.2, 10.2 and this Section. It may be appropriate to summarize them here:

a) the metal itself: the relative activity for the homogeneous CO hydrogenation appears to be Rh > Ru > Co > all other group VIII metals;
b) ligand influences on the insertion mode of the aldehyde(s): donor ligands favoring the formation of alkoxy; acceptor ligands favoring hydroxyalkyl intermediates;
c) ligand influences on the migratory insertion of CO: donor ligand additives favor this step;
d) pressure: very high pressure is necessary if appreciable amounts of glycol and higher polyols are wanted.

10.4 Conclusion

The homogeneous hydrogenation of CO requires, in general, relatively high pressure. Several reasons have been mentioned. The formation of the catalytically active hydridocarbonyl species and/or its stability in solution at the usually high reaction temperature, require pressure. This appears to be the major reason in hydroformylation, as born out by Wilkinson's catalyst, $HRh(CO)L_3$ $(L = P(C_6H_5)_3)$, which is sufficiently active to work at room temperature, where it is stable. The other reason is the unfavorable thermodynamic situation in the polyol synthesis, which has to be counterbalanced by pressure.

An important mechanistic feature of the homogeneous reactions is the availability of intermolecular reductive elimination of intermediates and products, involving the participation of hydridometalcarbonyl complexes.

10.5 References

van Boven M, Alemdaroglu NH, Penninger JML (1975) Ind Eng Chem Prod Res Dev 14:259

Breslow DS, Heck RF (1960) Chem & Ind 467

Brown CK, Wilkinson G (1970) J Chem Soc (A) 2753

Chan ASC, Carroll WE, Willis DE (1983) J Mol Catal 19:377

Clark HC, Davies JA (1981) J Organometal Chem 213:503

Collman JP, Hegedus LS (1980) Principles and Applications of Organotransition Metal Chemistry. University Science Books, Mill Valley, California

Cotton FA, Wilkinson G (1980) Advanced Inorganic Chemistry. Interscience, 4th ed.

Dombek BD (1980) J Amer Chem Soc 102:6855

Dombek BD (1981) J Amer Chem Soc 103:6508

Fahey DR (1981) J Amer Chem Soc 103:136

Falbe J (1967) Synthesen mit Kohlenmonoxyd. Springer, Berlin and New York

Falbe J (ed) (1980) New Syntheses with Carbon Monoxide. Springer, Heidelberg, New York, Chapter 4.2

Feder HM, Rathke JW (1980) Ann NY Acad Sci 298:45

Heck RF, Breslow DS (1961) J Amer Chem Soc 83:4023

Henrici-Olivé G, Olivé S (1977) Coordination and Catalysis. Verlag Chemie, Weinheim, New York

Henrici-Olivé G, Olivé S (1977/78) J Mol Catal 3:443

Ichikawa M (1979) J Catal 59:67

Keim W, Berger M, Schlupp J (1980) J Catal 61:359

Kiennemann A, Jenner G, Bagherzadah E, Deluzarche A (1982) Ind Eng Chem Prod Res Dev 21:418

Knifton JF (1982) J Catal 76:101

Knifton JF (1983a) J Chem Soc Chem Commun 729

Knifton JF (1983b) J Catal 79:147

Markó L (1961) Ungarische Mineralöl- und Erdgas-Versuchsanstalt 2:228

Markó L (1974) in: Ugo R (ed) Aspects of Homogeneous Catalysis. Riedel, Dordrecht, Holland

Martin, JT, Baird MC (1983) Organometallics 2:1073

Matsumoto M, Tamura M (1983) J Mol Catal 19:365
Moffat AJ (1970) J Catal 19:322
Orchin M, Rupilius W (1972) Cat Rev 6:85
Parker DG, Pearce R, Prest DW (1982) J Chem Soc Chem Commun 1193
Paulik FE (1972) Cat Rev 6:49
Pino P (1980) J Organometal Chem 200:223
Pruett RL (1977) Ann NY Acad Sci 295:239
Roelen O (1938) DRP 849,548 (see also Angew Chem 60:62 (1948))
Roth J, Orchin MJ (1979) J Organometal Chem 172:C27
Saus A, Phu TN, Mirbach MJ, Mirbach MF (1983) J Mol Catal 18:117
Sisak A, Ungváry R, Markó L (1983) Organometallics 2:1244
Spencer A (1980) J Organometal Chem 194:113
USP (1950) 2,534,018 (to DuPont de Nemours)
USP (1953) 2,636,046 (to DuPont de Nemours)

11 Methanol as Raw Material

Methanol is synthesized from CO and H_2 in large quantities at *low price* and with a high degree of *purity* (Danner, 1970; see also Sect. 8.4). This is the reason for the large amount of work towards syntheses starting with methanol, and aiming at the same or similar compounds obtainable from the direct hydrogenation of CO. In this Chapter, the most important of these methanol based processes will be discussed.

11.1 Carbonylation of Methanol (Acetic Acid Synthesis)

The synthesis of acetic acid from methanol according to

$$CH_3OH + CO \xrightarrow{\text{I}} CH_3COOH \qquad (11.1)$$

is a homogeneous process of considerable industrial importance. As in hydroformylation, cobalt and rhodium salts or carbonyl complexes are the most important catalyst precursors, although ruthenium and iridium have also been recognized as being capable of catalyzing the reaction (Forster and Singleton, 1982). A "halogen promoter" (halogen or halide), preferably iodide, is indispensable. The older cobalt process (Hohenschutz et al., 1966) operates at 210 °C and 500 atm CO; about 90% of the methanol is converted to acetic acid. The more recently developed *rhodium catalyst* (Roth et al., 1971) can be used at much lower pressure and provides reaction rates essentially independent of the CO pressure in the range between 13 and 120 atm. A convenient operating temperature is 175 °C. The formation of acetic acid proceeds with a degree of selectivity (99%) which is remarkable even for a homogeneous catalytic process. Worldwide, 90% of the new production capacity brought onstream since 1973 has been based on the Monsanto rhodium catalyzed process (Forster and Singleton, 1982).

We shall consider the rhodium catalyzed reaction in more detail, following the work of Forster (1975, 1976). The catalyst precursor comprises two components, a soluble rhodium compound and an iodide promoter. Invariant reaction rates have been obtained utilizing a variety of rhodium compounds (e.g. $RhCl_3 \cdot 3H_2O$, Rh_2O_3, $[R_4As]^+[Rh(CO)_2I_2]^-$), and any of several promoter components (e.g. aqueous HI, CH_3I, I_2), under comparable operation conditions. This indicates that, ultimately, the same active species is formed, which is assumed to be the anionic Rh(I) complex $[Rh(CO)_2I_2]^-$. The most convenient catalyst system consists of the commonly available $RhCl_3 \cdot 3H_2O$ and aqueous hydrogen iodide, in a water/acetic acid mixed solvent.

Although the net reaction is described by Eq. (11.1), several equilibria are involved during the course of the reaction:

$$2 CH_3OH \rightleftarrows CH_3OCH_3 + H_2O$$

$$CH_3OH + CH_3COOH \rightleftarrows CH_3COOCH_3 + H_2O$$

$$CH_3OH + HI \rightleftarrows CH_3I + H_2O.$$

It has been established that methyl iodide attains a steady state concentration very rapidly, and remains constant during most of the reaction. The other two equilibria are shifted during the reaction and ultimately all of the intermediates are converted into acetic acid. Kinetic studies resulted in the following rate law:

$$\text{rate} = k\, [CO]^0\, [CH_3OH]^0\, [Rh]\, [I^-]. \tag{11.2}$$

Investigation of the mechanism using a combination of techniques including in situ spectroscopy and stoichiometric transformation with species established in the reaction, led to the following proposal for the reaction pathway:

$$CH_3OH + HI \rightleftarrows CH_3I + H_2O \tag{11.3}$$

$$[Rh(CO)_2I_2]^- + CH_3I \rightarrow [CH_3Rh(CO)_2I_3]^- \tag{11.4}$$

$$[CH_3Rh(CO)_2I_3]^- + CO \rightarrow [CH_3\overset{\|}{\underset{O}{C}}Rh(CO)_2I_3]^- \tag{11.5}$$

$$[CH_3\overset{\|}{\underset{O}{C}}Rh(CO)_2I_3]^- \rightarrow CH_3COI + [Rh(CO)_2I_2]^- \tag{11.6}$$

$$CH_3COI + H_2O \rightarrow CH_3COOH + HI. \tag{11.7}$$

Methyl iodide formed in equilibrium (11.3) is oxidatively added to the square planar Rh(I) complex (Eq. 11.4). The resulting methyl rhodium complex must be very short-lived, since spectroscopic monitoring of the reaction between the square planar catalyst and CH_3I shows only the formation of the acetyl complex, generated by insertion of CO into the methyl-rhodium bond

(Eq. 11.5). Reductive elimination of CH_3COI (Eq. 11.6) reestablishes the catalyst. Hydrolysis of the acetyl iodide produces acetic acid and HI (Eq. 11.7).

The postulated reaction mechanism is consistent with the kinetic result (Eq. 11.2), if one assumes that the oxidative addition of methyl iodide is rate determining, and all other steps occur rapidly.

Thus, the role of the iodide in the system has been clearly related to the generation of methyl iodide, which in turn makes possible the formation of the metal-carbon bond by oxidative addition. This view is corroborated by the fact that iodide sources incapable of generating methyl iodide in the system in significant quantities (e.g. alkali metal iodides) do not promote the reaction (Forster and Singleton, 1982).

Heterogeneization of the rhodium catalyst has been undertaken by several authors (see e.g. Nefedov et al., 1976; Christensen and Scurrell, 1977). Zeolite supported Rh catalysts have shown high activity and selectivity for methanol carbonylation. At 340 °C and normal pressure, methyl acetate is the main product (> 90%). Nickel deposited on active carbon has been reported to catalyze the vapor phase carbonylation of methanol (Inui et al., 1981; Fujimoto et al., 1982). At 300 °C, 11 atm (CO/methanol = 5) in the presence of iodide promoter, up to 95% acetic acid was obtained (Fujimoto et al., 1983). In contrast to the solution process, the initial product is methyl acetate, the hydrolysis of which is assumed to result in the final product acetic acid. At T < 250 °C, methyl acetate and dimethyl ether are the major products. The process appears interesting since it uses Ni instead of the costly and rare rhodium.

11.2 Methanol Homologation

The first reports of the cobalt-catalyzed hydrocarbonylation (or homologation) of methanol to give ethanol and/or acetaldehyde, date back to 1941 (DRP, 1941). After a period of relative inactivity, the reaction has become the subject of an increasing number of studies recently. This may have been sparked by estimates of economic analysts that this might be a potentially interesting route to coal-based ethylene (e.g. Haggin, 1981).

The homogeneously catalyzed reaction is generally carried out at 200−220 °C and 250−300 atm, with a H_2/CO ratio of 1−1.5. Formally, it may be written as:

$$CH_3OH + CO + H_2 \rightarrow CH_3CHO + H_2O \qquad (11.8)$$

$$CH_3CHO + H_2 \quad \rightarrow CH_3CH_2OH. \qquad (11.9)$$

However, a number of byproducts are generally formed: acetals, esters, ethers, and acetaldehyde condensation products. Limited selectivity appears to have prevented industrial application so far.

The reaction is significantly promoted by iodine or iodides; donor ligands, such as phosphines enhance the stability and selectivity of the catalyst system (Röper et al., 1982, and references therein). Mechanistic studies have been made with the catalyst system $Co(OAc)_2 \cdot 4H_2O$, PH_3P, and aqueous HI. Best results were obtained with a cobalt/phosphine/iodide ratio of $1:2:2$. Deuterated methanol submitted to the homologation reaction at 300 atm CO/H_2 ($1:1$), 200 °C led to the following products (Röper et al., 1982):

$$CD_3OD + CO + H_2 \rightarrow CD_3CHO + CD_3CH(OCD_3)_2 +$$

$$+ CD_3CH_2OH + CD_3COOCD_3.$$

This result showed unequivocally that the CD_3 group of methanol remains intact during the reaction, and invalidated early suggestions that the reaction might involve carbene (Ziesecke, 1952). A mechanism very closely related to Forster's acetic acid synthesis was suggested by Röper et al., involving the formation of CH_3I, its oxidative addition to a square planar Co(I) complex, CO insertion into the $Co-CH_3$ bond, and reductive elimination of acetaldehyde, followed by secondary reactions of the latter:

$$CH_3OH + HI \quad \rightleftarrows \quad CH_3I + H_2O \tag{11.10}$$

$$Co(CO)L_3 + CH_3I \quad \rightarrow \quad CH_3Co(CO)IL_3 \tag{11.11}$$

$$CH_3Co(CO)IL_3 + CO \rightarrow CH_3\overset{\|}{\underset{O}{C}}Co(CO)IL_3 \tag{11.12}$$

$$CH_3\overset{\|}{\underset{O}{C}}Co(CO)IL_3 \xrightarrow{\text{hydrogen}} CH_3CHO + \text{catalyst} . \tag{11.13}$$

The exact composition of the active species is not known; Röper et al. suggested $Co(CO)IL_2$. The last step (Eq. 11.13) was assumed to involve loss of I_2, in order to accommodate oxidative addition of H_2 to the Co center. However, based on the chemistry of phosphine modified cobalt catalysts for the hydroformylation of olefins (see e.g. Henrici-Olivé and Olivé, 1977) one may assume that $HCo(CO)_nL_m$ species are present under the reaction conditions. (They may even be the catalytically active species in Eq. 11.11.) As in hydroformylation, we are then left with the question whether the reductive elimination (step 11.13) takes place intermolecularly with a cobalt hydrido species (cf. step 5 in Scheme 10.1), or via oxidative addition of H_2, as assumed by Röper et al. In view of the evidence for intermolecular reductive elimination in homogeneous systems, as exposed in Chap. 10, this path may well be considered as an alternative in this system too.

For phosphine-free Co catalysis of the methanol homologation reaction a different mechanism has been suggested, involving protonation of CH_3OH by the acidic $HCo(CO)_4$ and subsequent dehydration for the formation of the key methyl ligand (Hecht and Kröper, 1948; Slocum, 1980):

$$HCo(CO)_4 + CH_3OH \rightarrow CH_3OH_2^+ + Co(CO)_4^- \xrightarrow{-H_2O} CH_3Co(CO)_4. \quad (11.14)$$

Although this particular sequence has not been demonstrated, e.g. by a stoichiometric reaction, there is some evidence that the mechanism might be different in the presence or absence of phosphine ligands: the phosphine containing system is not promoted by ionic iodides such as NaI or KI; these substances are, however, good activators for the phosphine-free system. This may indicate a nucleophilic attack of I^- in the latter system, which would be prevented by the presence of the donor ligand in the former.

Interestingly, the phosphine free cobalt system, in the presence of KI plus CH_3I, is highly selective towards acetaldehyde ($> 90\%$), at 250 atm, 205 °C (Gauthier-Lafaye et al., 1982). The hydrogenation of the aldehyde is almost completely inhibited.

The other extreme, high selectivity towards ethanol, can be obtained with mixed Co/Ru catalysts (Doyle, 1983). Ruthenium catalysts without cobalt, are very sluggish in this reaction, even in the presence of phosphine ligands and iodide promoters. They are, however, excellent hydrogenation catalysts. It is therefore assumed that the two metal centers operate separately, Co providing acetaldehyde, and Ru hydrogenating it to ethanol. Up to 77% selectivity to ethanol has been obtained with a catalyst system containing $Co_2(CO)_8$, CH_3I and $(\eta^5-C_5H_5)Ru[P(C_6H_5)_3]Cl$.

Frequently, small amounts of propanol are also present in those systems giving a large yield of ethanol. They could be formed by further homologation of ethanol, although this must be a very slow reaction, since the formation of ethyl iodide from ethanol and HI is much less favorable than equilibrium (11.10), and moreover, C_2H_5I is very sluggish in oxidative addition reactions, as compared with CH_3I. Furthermore, ^{14}C-tracer evidence has been reported by Burns (1955) that propanol might be formed by hydroformylation of ethylene resulting from the dehydration of C_2H_5OH.

High selectivity to ethanol (up to 72%) has also been found with mixed $Fe(CO)_5/Mn_2(CO)_{10}$ catalysts in methanol/amine solutions (Chen et al., 1982). Addition of methyl iodide increases the rate of methanol conversion, but has no effect on the selectivity. The mechanism needs further elucidation.

Finally, it may be noted that homologation is not restricted to methanol. Methyl ether has been transformed to methyl acetate and acetic acid (Braca et al., 1979), and methyl acetate to ethyl acetate (Braca et al., 1982), with iodide promoted ruthenium catalysts, at 200 °C and 250 atm.

11.3 Hydrocarbons from Methanol Dehydration and Condensation

The conversion of methanol into hydrocarbons with strong dehydration agents such as P_2O_5 has been known for some time (Pearson, 1974). At 190 °C mostly branched hydrocarbons are formed. The great problem is the lack of selectivity, about 200 compounds being produced. Methanol on bulk ZnI, at 200 °C and 13.6 atm N_2, was also reported to produce highly branched hydrocarbons, 42% of which were in the gasoline, 56% in the gas-oil range (Kim et al., 1978).

The interest of this type of reaction started to boom when Mobil announced that the use of a particular zeolite, called ZSM-5, permits the selective formation of hydrocarbons, the maximum size of the product molecules being limited to the C_9-C_{10} range (Meisel et al., 1976). The reaction products are mainly branched chain and aromatic hydrocarbons, ideal for high octane fuel. The process was ready for pilot plant trials in 1977 (Lee et al., 1978). A major problem was heat dissipation, because the reaction is highly exothermic (ca. 1.7 MJ per kg methanol transformed).

The potential of *zeolites* for shape selectivity has been recognized many years ago by Weisz et al. (1962). Some typical examples have been discussed in Sect. 6.2.3. A breakthrough in zeolite catalysis began when it was realized that by addition of appropriate cations to the zeolite synthesis mixture, completely new structures could be obtained. One of the most significant developments was the synthesis by Mobil of the zeolite ZSM-5 (Chang and Silvestry, 1977). As all zeolites, ZSM-5 is made up of linked SiO_4 and AlO_4 tetrahedra. The structure consists of a system of two intersecting channels (diameter $\simeq 0.55$ nm) going along the a and b axes; those along the b axis are straight and circular, whereas those parallel to a are sinusoidal and of elliptical cross-section (Thomas and Millward, 1982). Figure 11.1 shows some structural details of ZSM-5. The access to the inner channel system is controlled by openings with diameters of the order of 0.55 nm, which is considerably smaller than the "windows" in the faujasite type zeolites (cf. Fig. 6.3).

Table 11.1 shows the mixture of hydrocarbons produced at various temperatures, amounting to the theoretical conversion of methanol to 44% hydrocarbons and 56% water according to

$$x\ CH_3OH\ \rightarrow\ (CH_2)_x + x\ H_2O. \tag{11.15}$$

Methane and ethane represent only 1−5%; the distribution of molecular weights is narrower than in the Fischer-Tropsch synthesis (Chap. 9), what may represent a potential advantage in future applications.

High selectivity to C_2-C_4 olefins (80%) has been obtained by Inui et al. (1983) with a tetramethylammonium modified zeolite at 400 °C. With seed crystals as crystallization nuclei, the shape and size distribution of the zeolite could be uniformly regulated, and catalyst performance for olefin selectivity,

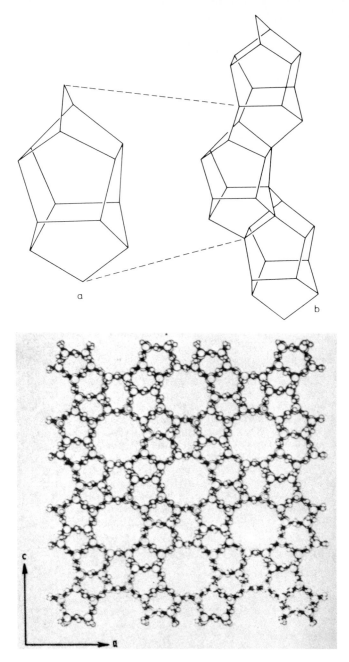

Fig. 11.1. (a) Segment of the structure of ZSM-5 showing connected 5-membered rings composed of linked tetrahedra (SiO_4 and AlO_4). Each connecting line represents an oxygen bridge. (b) Chains are made up by linking the units shown in (a). (c) Chains are linked together to form a very regular tridimensional network (Thomas and Millward, 1982). Reproduced by permission of The Royal Society of Chemistry

Table 11.1. Conversion of Methanol to Hydrocarbons over ZSM-5 Zeolite (Kaeding and Butter, 1980).

Reaction Conditions [a]			
Temperature (°C)	400	450	500
Conversion (wt%) [b]	100	100	100
Hydrocarbons	Weight percentage		
Methane	0.7	2.6	4.1
Ethane	0.2	0.9	1.2
Propane	6.9	11.4	14.9
Butanes	20.1	19.8	19.3
Total C_2-C_4 paraffins	27.2	32.1	35.4
Ethylene	3.4	6.4	6.1
Propylene	3.9	6.4	7.3
Butylenes	9.6	7.4	6.4
Total C_2-C_4 olefins	16.9	20.2	20.5
C_5-C_{10} aliphatics	39.9	23.0	15.6
Aromatics [c]	15.3	22.1	24.5
Total	100.0	100.0	100.0
C_2-C_4 olefin/paraffin ratio	0.62	0.63	0.56

[a] Atmospheric pressure; weight hourly space velocity = 1.
[b] To hydrocarbons and water.
[c] Primarily benzene, toluene and xylenes.

as well as catalyst life were improved. Predominantly C_2-C_4 alkenes (> 80%) were also formed on a different type of zeolite, viz. barium ion-exchanged mordenite, at 425 °C (Itoh et al., 1981).

Pichler and Ziesecke (1949) had reported on the formation of highly branched hydrocarbons, starting directly from $CO + H_2$, using ThO_2 or ThO_2/Al_2O_3 as catalyst. This idea has been pursued by Kieffer et al. (1983), who used the rare earth metal oxides La_2O_3 and Dy_2O_3 as catalysts (400 atm, 410−475 °C). The authors could show that the primary product is methanol, which is further transformed in a Mobil type process to mainly branched C_4. However, conversion is low. The addition of Pd, a good catalyst for the formation of methanol from $CO + H_2$ (see Sect. 8.3) to the rare earth metal oxide catalyst led to improved conversion, but also to increased methane evolution.

The mechanism of the Mobil process of converting methanol to hydrocarbons is far from being fully elucidated. However, some aspects of the overall picture are known with certainty. Methyl ether and water are formed initially and at high rates; operating at relatively low temperature (250 °C) the ether could be produced as the major product in excellent yield (Kaeding and Butter, 1980). If the acidic sites of the zeolites are blocked with pyridine, no hydrocarbons are formed; dimethyl ether is the only product at 239 °C (Ono and Mori, 1981). Thus, the acid sites appear to be essential for the formation of hydrocarbons. The rate of formation of dimethyl ether was reduced but not inhibited by the pyridine; after 47 h the ether yield was 99.9% and only a trace of methane was present. This may imply that the formation of dimethyl ether proceeds on weak acid sites, which do not retain adsorbed pyridine molecules at 239 °C, while the hydrocarbon formation requires stronger acid sites.

Ono and Mori (1981) studied the conversion of methanol in a closed circulating system, instead of the usual flow reactor. This procedure permitted them to carry out the reaction at lower temperature, and to examine every stage in "slow-motion". Although the product distribution was certainly not quite the same as at 400−500 °C (the average molecular weight was lower), the basic findings are assumed to be transferable to the higher temperature process. At 221 °C dimethyl ether was the main product during the first 8 h; only very small amounts of hydrocarbons were detected. After 12 h the hydrocarbon yield increased abruptly, reaching 80% in 18 h. Thereafter the reaction proceeded in steady state. This behavior indicates an autocatalytic process. The induction period could be drastically shortened by the addition of ethylene or cis-butene, showing that the autocatalysis is caused by the presence of product olefins. Apparently, the reaction of methanol with olefins proceeds much faster than the formation of incipient olefins from methanol (or dimethyl ether).

The overall reaction has been summarized by Chang and Chu (1982):

$$2\,CH_3OH \xrightarrow{-H_2O} CH_3OCH_3 \xrightarrow{-H_2O} \text{olefins} \rightarrow \text{isoalkanes} \rightarrow \text{aromatics}$$

$$(11.16)$$

The first step, ether formation appears straightforward; the reactions of olefins over acidic zeolites leading to isoalkanes and aromatics are also well understood and explainable by classical carbenium ion mechanisms (Poutsma, 1976). But the mechanism of initial C−C bond formation from methanol is unknown, and has been the subject of much speculation. At least six different pathways have been suggested by various authors, invoking surface alkoxides, carbenes, oxonium ions, carbenium ions, free radicals, or pentacoordinated carbon as intermediates (Chang and Chu, 1982, and references therein). Perot et al. (1982) tried to verify by ^{13}C labeling a mechanism suggested by van den Berg et al. (1980) and involving alkylation of dimethyl ether to trimethyloxonium cation. But Fărcaşiu (1983) showed that the results of the labeling experiments are compatible with at least three of the suggested mechanisms.

Chang and Chu (1982, 1983), although convinced that carbene-like C_1 intermediates are present, and may be involved in the synthesis, state that the precise details of the involvement remain a mystery. There seems to be much opportunity for future scientific work.

11.4 Conclusion

The methanol-based syntheses will probably play an important role when, some day in the future, the use of coal-based synthesis gas instead of oil will become imperative.

With the acetic acid process, the future has already begun, the methanol based synthesis is worldwide well established in Industry. The polyol synthesis (especially ethylene glycol) still has to overcome problems of product selectivity; also the high pressure required may be a handicap. The Mobil process, though far from being understood from a mechanistic point of view, is well advanced technically, a pilot plant for the conversion of 4 barrels of methanol per day to gasoline has been successfully operated (Lee et al., 1978). Although not based on methanol, but on natural gas as feestock, a gasoline plant with ZSM-5 technology is under construction in New Zealand (Whan, 1981).

11.5 References

Braca G, Sbrana G, Valentini G, Andrich G, Gregorio G (1979) in: Tsutsui M (ed) Fundamental Research in Homogeneous Catalysis. Plenum Press, New York, Vol. 3, p. 221
Braca G, Sbrana G, Valentini G, Cini M (1982) J Mol Catal 17:323
Burns GR (1955) J Amer Chem Soc 77:6615
Chang CD, Chu CTW (1982) J Catal 74:203
Chang CD, Chu CTW (1983) J Catal 79:244
Chang CD, Silvestri AJ (1977) J Catal 47:249
Chen MJ, Feder HM, Rathke JW (1982) J Mol Catal 17:331
Christensen BC, Scurrell MS (1977) J Chem Soc Faraday Trans 2036
Danner CA (ed) (1970) Methanol Technology and Economics. Chem Eng Progr Symp Ser No 98:66
Doyle G (1983) J Mol Catal 18:251 (and references therein)
DRP (1941) 867,849 (to BASF)
Fǎrcaşiu D (1983) J Catal 82:252
Forster D (1975) J Amer Chem Soc 97:951
Forster D (1976) J Amer Chem Soc 98:846
Forster D, Singleton TC (1982) J Mol Catal 17:299 (and references therein)
Fujimoto K, Shikada T, Omata K, Tominaga K (1982) Ind Eng Chem Prod Res Dev 21:429
Fujimoto K, Omata K, Shikada T, Tominaga H (1983) Ind Eng Chem Prod Res Dev 22:436
Gauthier-Lafaye J, Perron R, Colleuille Y (1982) J Mol Catal 17:339
Haggin J (1981) Chem Eng News 59:(20), 52; see also Chem Eng News 60:(13), 28 (1982)
Hecht O, Kröper H (1948) Naturforschung und Medizin in Deutschland (FIAT Rev., German Sci.) 1939–46, Vol 36, Pt I, p 136
Henrici-Olivé G, Olivé S (1977) Coordination and Catalysis. Verlag Chemie, Weinheim, New York, Chap. 10.2
Hohenschutz H, von Kutepow N, Himmele W (1966) Hydrocarbon Processing 45:141
Inui T, Matsuda H, Takegami Y (1981) J Chem Soc Chem Commun 906
Inui T, Ishihara T, Morinaga N, Takeuchi G, Matsuda H, Takegami Y (1983) Ind Eng Chem Prod Res Dev 22:26
Itoh H, Hattori T, Murakami Y (1981) J Chem Soc Chem Commun 1091
Kaeding WD, Butter SA (1980) J Catal 61:155

Kieffer R, Varela J, Deluzarche A (1983) J Chem Soc Chem Commun 763

Kim L, Wald MM, Brandenberger SG (1978) J Org Chem 43:3432

Lee W, Maziuk J, Thiemann WK (1978) 26 DGMK-Haupttagung, Berlin, Preprints, p 675

Meisel SL, Mc Cullough JP, Lechthaler CH, Weisz PB (1976) CHEMTECH 86

Nefedov BK, Sergeeva NS, Zheva TV, Shutkina EM, Eidus YaT (1976) Izvest Akad Nauk SSSR, Ser Khim 582

Ono Y, Mori T (1981) J Chem Soc Faraday Trans I 77:2209

Pearson DE (1974) J Chem Soc Chem Commun 397

Perot G, Cormerais FX, Guismet M (1982) J Chem Res (S), 58

Pichler H, Ziesecke KH (1949) Brennstoff-Chem 30:13, 60, 81, 313

Poutsma ML (1976) in Zeolite Chemistry and Catalysis (Rabo JA, ed), ACS Monograph 171: Amer Chem Soc Washington DC, p 437

Röper M, Loevenich H, Korff J (1982) J Mol Catal 17:315

Roth JF, Craddock JH, Hershman A, Paulik FE (1971) CHEMTECH 600

Slocum DW (1980) in Johns WH (ed) Catalysis in Organic Chemistry, Academic Press, New York, p 245

Thomas JM, Millward GR (1982) J Chem Soc Chem Commun 1380 (and references therein)

van den Berg JP, Wolthuizen JP, van Hooff JHC (1980) in Rees LV (ed) Proc. 5th Internat. Conf. on Zeolites. Heyden, London, p 649

Weisz PB, Frilette VJ, Maatman RW, Mower EB (1962) J Catal 1:307

Whan DA (1981) Chem & Ind 532

Ziesecke KH (1952) Brennstoff Chem 33:385

12 Attempt of a Unified View

In this Chapter we try to unify the mechanistic details we have been working through in the course of the book, to a general picture of the catalyzed hydrogenation of CO. It should be born in mind that neither the mechanistic details referred to, nor the intended unified picture are proven. They are the best guess, taking into account presently known experimental data, and avoiding to violate known principles of the reaction patterns of transition metal centers.

Let us first summarize some facts that have to be considered. The *homogeneous* syntheses generally require high pressure (methanol > 300 atm; glycol > 480 atm and up to 1500 atm; hydroformylation > 100 atm). The CO bond does not open, no water is evolved. The H_2/CO ratio is in most cases $1:1$. In the hydroformylation case it has been made highly probable that the reduction of the acyl ligand takes place by intermolecular reductive elimination, with a metal hydride species (cf. Sect. 10.1);

$$RCM(CO)_x + H-M(CO)_y \ \rightarrow \ RCH + M_2(CO)_{x+y}. \qquad (12.1)$$
$$\|\| $$
$$OO$$

The *heterogeneous* Fischer-Tropsch synthesis of hydrocarbons and/or alcohols takes place at normal to medium pressure (5–30 atm). The CO bond opens in every growth step, H_2O is evolved. The H_2/CO ratio is generally > 1. No unambiguously homogeneous Fischer-Tropsch system has been found.

The industrial methanol synthesis (Cr/Zn: 250–350 atm; Cu/ZnO: 50–100 atm) is heterogeneous, yet no chain growth takes place; the CO bond remains unopened and no water is evolved.

We want to suggest that under the high CO pressure conditions of the *homogeneous* syntheses, hydrogen has little chance to compete with CO for coordination sites (note that two such sites in cis position must be available for the oxidative addition of the H_2 molecule). As a consequence, all hydrogenation steps in the homogeneous mechanisms (Schemes 8.1, 10.1, 10.3) would be in fact intermolecular reductive eliminations of the type shown in Eq. 12.1). The existence of intermolecular reductive elimination in solution is well documented (cf. Sect. 5.2.2); it requires only one vacant site which might be the one vacated by preceding CO insertion reaction.

Certain quantitative aspects of the relative rates of the reactions of acyl- or alkyl-complexes with a metal hydride and H_2 have been studied recently by Markó and his coworkers (Ungváry and Markó, 1983; Hoff et al., 1984). Using the relatively stable model complexes $C_2H_5O(CO)Co(CO)_4$ and $C_2H_5O(CO)CH_2Co(CO)_4$, the authors investigated the kinetics of the following homogeneous reactions (n-octane solution, 25 °C):

$$RCo(CO)_4 \; \underset{k_{-1}}{\overset{k_1}{\rightleftarrows}} \; RCo(CO)_3 + CO$$

$$RCo(CO)_3 \begin{cases} \xrightarrow[k_2]{HCo(CO)_4} RH + Co_2(CO)_8 \\ \xrightarrow[k_3]{H_2} RH + HCo(CO)_4 \end{cases}$$

It was found that in the case of the acyl-complex ($R = C_2H_5O(CO)$) the rate constant for the reaction with $HCo(CO)_4$ is an order of magnitude higher than that of the reaction with H_2. For the alkyl-complex ($R = C_2H_5O(CO)CH_2$) the rate constant is even two orders of magnitude higher for the hydride than for H_2. Although the investigated model compounds do not necessarily represent exactly the active species in catalytic cycles, the trend is remarkable, and evidently corroborates the hypothesis that not only acyl-, but also alkyl-complexes react predominantly with hydride complexes, as opposed to H_2, in homogeneous media.

The assumption of intermolecular reductive elimination as the most probable hydrogenation step immediately makes plausible the absence of chain growth in homogeneous systems: since at no time a hydride and a hydroxyalkyl ligand are present at one and the same center, water elimination (step 7 in Scheme 9.1, Section 9.5.3) cannot take place. Carbon monoxide insertion steps, on the other hand, are no problem, since no extra coordination site is necessary: ligand CO is inserted into a M-ligand bond.

In Scheme 12.1, all reactions possible along these lines in a homogeneous system are summarized. As noted in the Introduction, "everything that can happen in a catalytic reaction will happen; selectivity is just a matter of relative rates" (Herman Bloch). As we have seen in Chaps. 8 and 10, selectivity is in fact a problem in most homogeneous CO/H_2 based syntheses. In some cases ligand influences on selectivity have been observed, either by orienting the insertion of formaldehyde (or higher aldehydes) towards intermediate *1* or *2* (or the corresponding higher intermediates *5* and *6*), or by favoring CO insertion over intermolecular reductive elimination (examples for both have been given in earlier Chapters). Very high pressure leads to chain growth (CO insertion), evidently because the intermolecular reductive elimination becomes more difficult (the necessary single free coordination site becomes less available, since CO blocks it at the higher pressure); consequently, the formation of intermediates *3* and *4*, and their further reactions become less favorable.

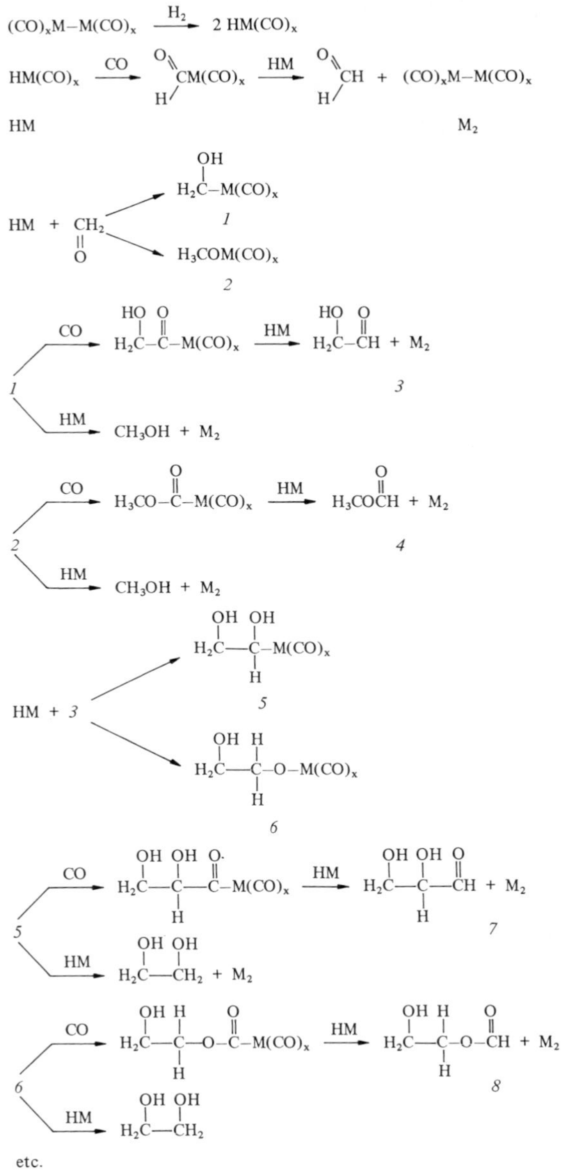

Scheme 12.1   etc.

A few additional remarks with regard to Scheme 12.1 may be appropriate. According to this Scheme, formaldehyde which is formed by the first intermolecular reductive elimination step of the Scheme, leaves the catalytic center, because no coordination site is available in the coordinatively saturated metal carbonyl cluster (e.g. $Co_2(CO)_8$). This is in contrast to heterogeneous systems (Scheme 9.1), where formaldehyde (and higher aldehydes) appear to remain attached to the catalyst center.

The declustering of the metal complex by hydrogen has been formulated for a binuclear cluster, as it is found up to high pressure in the case of $Co_2(CO)_8$. In other cases higher clusters may be involved. In the case of the Rh catalyzed polyalcohol synthesis, for instance, pressure dependent equilibria between several polynuclear rhodium carbonyl clusters appear to exist at the very high pressures used for the synthesis (Vidal and Walker, 1980; Fahey, 1981). Whether under these conditions declustering by hydrogen leads also to mononuclear species as in the case of cobalt has to remain open. In any case, hydrido complexes of the general form $HM_x(CO)_y$ can be assumed to be formed.

The declustering by H_2 (first reaction in Scheme 12.1) is known to be a slow reaction (see Sect. 10.1, Eq. (10.2), and also Sisak et al., 1983); it may well be the slowest step of the whole sequence. The rates of the homogeneous syntheses show a linear dependence of the hydrogen partial pressure (see, e.g. Henrici-Olivé and Olivé (1977) for hydroformylation of olefins; Feder and Rathke (1980) for the homogeneous methanol synthesis); if Scheme 12.1 is correct, declustering by H_2 is the only step where hydrogen pressure comes to bear directly. Each intermolecular reductive elimination step rebuilds the cluster compound.

Successive CO insertions have not been included in Scheme 12.1: they have never been observed (cf. Sect. 5.3.4). Formates such as *4* and *8* do not react further (the corresponding products have not been reported).

Scheme 12.1 implies that the homogeneous syntheses (hydroformylation, methanol and polyol syntheses) are confined to the liquid medium which warrants the necessary mobility of the hydridometal carbonyl species. In fact, when "heterogenized" catalysts have been found active, it was a suspension of catalyst in a liquid medium, and leaching of metal carbonyl was observed; hence the mobile $HM(CO)_x$ species were actually available (cf. Sect. 10.1).

In the *heterogeneous* Fischer-Tropsch systems, intermolecular reductive elimination is not available, because all metal centers are fixed on a surface. The only possibility of activation of H_2 is its oxidative addition to metal centers. The relatively low overall pressure and the generally used ratio $H_2/CO > 1$ improve the opportunity for hydrogen to compete with CO as a ligand. Most probably, these oxidative additions (two of them are required for each C addition, see Scheme 9.1) are slow, as indicated by the fact that the rate increases linearly with P_{H_2}. The simultaneous presence of hydrogen and hydroxyalkyl ligand at the same active center permits elimination of water (step 7, Scheme 9.1), and hence chain growth. This, then, accounts for the fact that only heterogeneous Fischer-Tropsch systems have been found. It may be interesting in this context that already in the early days of the Fischer-Tropsch synthesis, Fischer and Küster (1933), when trying to carry out the synthesis with a cobalt catalyst suspended in paraffin oil, found a considerable increase of oxygenates. Apparently, cobalt leaching into the liquid phase as $HCo(CO)_4$ had transformed the system into a mixed heterogeneous/homogeneous catalyst. Although the main products of a typical FT synthesis are hydrocarbons, there are always small amounts of alcohols present (see Table 9.1), and they are also primary products (Sect. 9.2.1). Scheme 9.1 suggests their formation by intramolecular reductive elimination of the α-hydroxyalkyl ligand. Based on the reactions taking place in homogeneous systems (Scheme 12.1), an alternative (or additional) pathway to alcohols in heterogeneous systems may be considered. Formaldehyde (and the higher aldehydes successively generated in the FT growth process) have, in principle, two possible modes of insertion into a metal−R bond (R = H or growing chain):

$$
\begin{array}{c}
\text{RCH} \\
\| \cdots \cdots \text{ M} \\
\text{O} \quad | \\
\quad \text{H}
\end{array}
\begin{array}{c}
\xrightarrow{a} \text{RCH}_2\text{O--M} \xrightarrow{\text{H}_2} \text{RCH}_2\overset{\overset{\displaystyle \text{H}}{|}}{\underset{\underset{\displaystyle \text{H}}{|}}{\text{O--M}}} \longrightarrow \text{RCH}_2\text{OH + M--H} \\[2em]
\xrightarrow{b} \text{RC--M} \longrightarrow \text{growth} \\
\quad\;\; | \\
\quad\;\; \text{OH}
\end{array}
\qquad (12.2)
$$

The generation of an alkoxy ligand (step a in Eq. 12.2) would prevent water elimination; the ligand would be bound to leave the active center by intramolecular reductive elimination as an alcohol. As in step 6, Scheme 9.1, the active center, M−H, is regenerated.

The relative rate of alcohol formation, as compared to that of β-H abstraction leading to olefins (step 15 in Scheme 9.1), depends on the metal itself, the support, promoters, pressure and temperature. Conditions have been found, where the alcohol formation predominates (see Sect. 9.3.2).

The molecular weight distribution of the product molecules depends solely on the probability for chain growth α (unless non-system-related restrictions are present, as for instance in zeolites):

$$ \alpha = \frac{r_p}{r_p + \Sigma\, r_{tr}} $$

where r_p is the rate of chain growth, and $\Sigma\, r_{tr}$ is the sum of transfer by reductive elimination leading to alcohol, and transfer by β-H abstraction, leading to olefin (see Sect. 9.3.1). A growing chain does not "know" by which of the two mechanisms it is going to be terminated. Hence, if the molecular weight distribution of the hydrocarbons and that of the alcohols would be determined separately, they should result in the same distribution pattern (same α) as the overall product.

In a freshly activated FT catalyst (activated generally by a reductive treatment at $T \geqq 500\,°C$ with H_2 or CO/H_2) is used for a FT synthesis, generally an induction period is observed, during which the rate of reaction slowly increases until it has reached its steady state. The following interpretation is offered within the framework of the presently suggested reaction scheme. After the reductive treatment under severe conditions, most of the catalyst surface is reduced to the metallic state. When the syngas mixture is admitted under the milder conditions of the FT synthesis, CO and hydrogen compete for the available coordination sites, with CO dominating. A certain percentage of the surface metal atoms will have hydrogen ligands. But only those few situated at especially exposed positions (at steps, kinks, surface irregularities, cf. Figs. 2.2, 6.2, 7.3) will be able to actually start chain growth, because only these metal centers have the sufficient number of coordination sites available. We suggest that the induction period represents the time necessary to provide *all potential catalyst sites* with hydrido ligands, in a slow equilibrium ligand exchange process. Once a catalytic center has initiated chain growth, it will be regenerated after each chain leaves the center, by either of the two chain transfer mechanisms mentioned above (predominantly by β-H abstraction, occasionally by alcohol formation, both regenerating a M−H bond). Steady state rate is reached when all possible catalyst sites are engaged in chain growth.

The industrial methanol synthesis is different in that it is heterogeneous, and yet no water is eliminated, but methanol is formed with very high selectivity. Evidently, only step *a* (Eq. 12.2, with $R = H$) is possible for the insertion of formaldehyde into the metal−H bond, in these systems. The catalytic centers are Cu(I) in the case of the low pressure process (see Chap. 8). However, ZnO is also present, and a heterolytic splitting of hydrogen has been suggested (Sect. 8.4, Eq. 8.10). Apparently, the hydride ligand at the transition metal center, resulting from this heterolytic splitting, has sufficient negative charge to attack coordinated formaldehyde exclusively at the positive end of its $C^{\delta+} \rightarrow O^{\delta-}$ dipole. Once the methoxy ligand is formed, the way to water elimination and further growth is blocked, methanol is the only major product. (The extraordinary synergistic effect of ZnO has been discussed in more detail in Sect. 8.4.)

Thus, it appears that Scheme 9.1, with the possible modification given in Eq. (12.2), together with Scheme 12.1 account for all major experimental results. As mentioned at the beginning of this Chapter, this does not mean that they are proven, but they represent a logical and plausible link between the various CO hydrogenation reactions.

One final "provocative" line of thoughts may be added. Formaldehyde has been found to be the key intermediate common to all mechanisms. Mother Nature also uses formaldehyde as an important intermediate for numerous biosyntheses. Of course, the "feedstock" is not the life threatening CO, but CO_2 from the atmosphere. In an atmosphere of $^{14}CO_2$, $^{14}CH_2O$ has been found in corn leaves, as an intermediate in the photosynthesis of glycol aldehyde and glycerin aldehyde (Szarvas and Pozsar, 1979). The way from CO_2 to formaldehyde seems to be not known exactly. Could photochemical reduction of CO_2, in the presence of a transition metal, provide carbon monoxide (see Hawecker et al., 1983), reacting to formaldehyde at the same metal center, in a way equal or similar to the first steps in Scheme 12.1? The further products (glycol aldehyde and glycerin aldehyde) are also present in the Scheme (compounds *3* and *7*, respectively). Evidently, the hydroxymethyl routes are favored by Nature, indicating a positively charged "hydrido" ligand at the active center, as one may expect in aqueous medium (see Table 2.2). The formation of a hydroxymethyl group from formaldehyde is also brought about by *clostridium acidi urici*, which synthesizes serine from glycine and formaldehyde, according to the net formula:

$$
\begin{array}{ccc}
CH_2NH_2 & & CH_2OH \\
| & & | \\
C=O & + CH_2O \rightarrow & CHNH_2 \\
| & & | \\
OH & & C=O \\
& & | \\
& & OH
\end{array}
$$

(Houghland and Beck, 1979). Cells of *proteus vulgaris* carry out the same synthesis, and this has even been the subject matter of a patent (Japan Kokai, 1977). Formaldehyde administered to rats was found in the methyl groups of choline, creatine and methionine (Rachele et al., 1964). Microorganisms are known which actually catalyze the water gas shift reaction, others which produce acetic acid or volatile fatty acids from CO_2 and H_2 (Gold et al., 1980).

Does Mother Nature use reaction paths similar to those treated in this book? Or else, if Nature has available other routes to the same or similar reaction products, can we be guided by Nature to more economic low pressure processes with cheaper raw materials? Questions to be answered by future generations of chemists, biochemists and bioengineers.

12.1 References

Fahey DR (1981) J Amer Chem Soc 103:136
Feder HM, Rathke JW (1980) Ann NY Acad Sci 333: 5
Fischer F, Küster H (1933) Brennstoff-Chem 14:3
Gold DS, Goldberg I, Cooney CL (1980) 180th ACS Ann Meeting, PETR. 9
Henrici-Olivé G, Olivé S (1977) Coordination and Catalysis. Verlag Chemie, Weinheim, New York, Chapter 10
Hoff C, Ungváry F, King RB, Markó L (1984) J Amer Chem Soc, in press
Houghland AE, Beck JV (1979) Microbios 24:151
Japan Kokai (1977) 99,288 (to Tanabe Seiyaku Co Ltd.)
Rachele JR, White AM, Grünewald H (1964) J Biol Chem 239:353
Sisak A, Ungváry F, Markó L (1983) Organometallics 2:1244
Szarvas T, Pozsar BI (1979) Proc Hung Ann Meet Biochem 19th, p 35
Ungváry F, Markó L (1983) Organometallics 2:1608
Vidal JL, Walker WE (1980) Inorg Chem 19:896

Subject Index

CATALYSIS

Science and Technology

Editors: **J. R. Anderson, M. Boudart**

Catalysis: Science and Technology is a multivolume, comprehensive reference work. Catalysis is a subject where science and technology are closed linked, and the present work covers both of these aspects in depth. In general terms, the scope of **Catalysis: Science and Technology** is limited to topics which are, to some extent at least, relevant to industrial processes. In this sense, the whole of heterogenous catalysis falls within its ambit, but biocatalytic processes which have no significance outside of biology are not included. However, ancillary subjects such as surface science, materials properties, and other fields of catalysis are given adequate treatment, but not to the extent of obscuring the central theme. **Catalysis: Science and Technology** thus has a rather different emphasis from normal review publications in the field of catalysis: This work concentrates more on important, established material, although at the same time providing as systematic presentation of relevant data. The opportunity has also been taken, where possible, to relate the specific details of a particular catalytic topic to established principles in chemistry, physics and engineering. **Catalysis: Science and Technology** is the only up-to-date comprehensive reference work that treats both the scientific and technological aspects of catalysis.

Volume 3

1982. 91 figures. X, 289 pages
ISBN 3-540-11634-6

E. E. Donath: History of Catalysis in Coal Liquefaction. –
G. K. Boreskov: Catalytic Activation of Dioxygen. –
M. A. Vannice: Catalytic Activation of Carbon Monoxide on Metal Surfaces. – *S. R. Morrison:* Chemisorption on Nonmetallic Surfaces. – *Z. Knor:* Chemisorption of Dihydrogen

Volume 1

1981. 107 figures. X, 309 pages
ISBN 3-540-10353-8

H. Heinemann: History of Industrial Catalysis. –
J. C. R. Turner: An Introduction to the Theory of Catalytic Reactors. – *A. Ozaki, K. Aika:* Catalytic Activation of Dinitrogen. – *M. E. Dry:* The Fischer-Tropsch Synthesis. –
J. H. Sinfelt: Catalytic Reforming of Hydrocarbons.

Springer-Verlag
Berlin
Heidelberg
New York
Tokyo

New Syntheses with Carbon Monoxide

Editor: **J. Falbe**

1980. 118 figures, 127 tables. XIV, 465 pages
(Reactivity and Structure, Volume 11)
ISBN 3-540-09674-4

Contents: *B. Cornils:* Hydroformylation. Oxo
Synthesis, Roelen Reaction. – *H. Bahrmann,
B. Cornils:* Homologation of Alcohols. –
A. Mullen: Carbonylations Catalyzed by Metal
Carbonyls-Reppe Reactions. – *C. D. Frohning:*
Hydrogenation of the Carbon Monoxide. –
H. Bahrmann: Koch Reactions. – *A. Mullen:* Ring
Closure Reactions with Carbon Monoxide.

The book is a report on developments in carbon
monoxide chemistry, emphasis being placed on
advances made in the seventies. Topics such as
hydroformylation, Fischer-Tropsch chemistry,
carbonylation, S. N. G., as well as ring closure
reactions with carbon monoxide are treated in
depth. Current mechanistic theories are critically
discussed. Themes are of current interest due to
revival of synthesis gas chemistry since the oil
crisis in 1973. The book recapitulates only very
briefly the period covered by Dr. Falbe's previous
book (**Carbon Monoxide in Organic Synthesis,**
Springer-Verlag, 1970), focussing on current
developments. It gives a detailed review of carbon
monoxide chemistry and will help to establish a
link between research (industrial and academic)
and application.

Springer-Verlag
Berlin
Heidelberg
New York
Tokyo